Wild Again

Wild Again

The Struggle to Save the Black-Footed Ferret

David Jachowski

UNIVERSITY OF CALIFORNIA PRESS
Berkeley · Los Angeles · London

University of California Press, one of the most distinguished university presses in the United States, enriches lives around the world by advancing scholarship in the humanities, social sciences, and natural sciences. Its activities are supported by the UC Press Foundation and by philanthropic contributions from individuals and institutions. For more information, visit www.ucpress.edu.

University of California Press
Berkeley and Los Angeles, California

University of California Press, Ltd.
London, England

Library of Congress Cataloging-in-Publication Data

Jachowski, David, 1977–
Wild again : the struggle to save the black-footed ferret / David Jachowski.
pages cm
Includes bibliographical references and index.
ISBN 978-0-520-28165-3 (cloth : alk. paper)
1. Black-footed ferret—Conservation—United States. I. Title.
QL737.C25J32 2014
599.76'629—dc23

2013026026

Manufactured in the United States of America

23 22 21 20 19 18 17 16 15 14
10 9 8 7 6 5 4 3 2 1

In keeping with a commitment to support environmentally responsible and sustainable printing practices, UC Press has printed this book on Natures Natural, a fiber that contains 30% post-consumer waste and meets the minimum requirements of ANSI/NISO Z39.48-992 (R 1997) (*Permanence of Paper*).

Contents

Prologue

For better or worse, this is a book about black-footed ferret conservation that was written for everyone. It is not intended to be a comprehensive, technical review of every aspect of efforts to recover the black-footed ferret from the brink of extinction, something a scholar would buy for their bookshelf but rarely use. This book is meant to be taken from the shelf to engage you, to be passed on, bent, folded and dog-eared. Take it on that next road trip to the Great Plains. Open it at a campground in Badlands National Park. Take it to the U.S.–Mexican border and crack the spine while sitting on the Chihuahua grasslands, allowing grains of prairie dust to sneak between the pages, pages that will be stained with coffee cup marks after late nights of searching for badgers, swift foxes, and perhaps even black-footed ferrets.

As a consequence, this book does not contain all viewpoints on all issues related to black-footed ferret recovery. This is a personal book of stories told from personal experiences and perspectives that collectively, and hopefully, reflect the core conservation message for this species. It is a summary of the lengths that individuals and society are willing to go to preserve an endangered species—not just any endangered species, but one that was once considered extinct, then rediscovered, and later extirpated from the wild. At its population low it was, at the very least, one of the rarest carnivores, and likely the rarest species, in existence.

The story of ferret recovery is an engaging one, perhaps one of the more remarkable conservation stories in the United States. There is no need for fiction or hyperbole in discussing the events related to the years of active ferret conservation. Their plight simply grabbed me from a young age and pulled me into a part of the world that I never thought I would learn to love, or for which I would live to fight so dearly.

For every walk of life, there is a message to be learned that helps describe where you are from. It is what makes a place special, what defines your values, what you take pride in. When traveling overseas, one learns there are few greater American icons that grip people like the "Wild West." And contrary to modern tourist propaganda, the Wild West is not in California, or the Rocky Mountains, or the bottom of the Grand Canyon, but in the center of the country—in the vastness of the Great Plains.

On the Great Plains, grasses dominate the landscape. And on those grasslands, patches of prairie dogs bring the prairie alive in increased plant and animal diversity. And on some of those prairie dog colonies, the presence of black-footed ferrets best symbolizes a healthy, biodiverse piece of ground—a locality likely complete with badgers, swift foxes, burrowing owls, mountain plovers, and ferruginous hawks, some of the prototypical representatives of the prairie.

Over the past thirty years, black-footed ferret rediscovery and subsequent conservation have marked a pivotal time in the history of the western United States. During this time new species were no longer being discovered at a rapid pace and attention turned more toward conservation. Species became protected by law through the Endangered Species Act, and the field of conservation biology was born. Ecosystem restoration was initiated on a scale whereby humans began to try to re-create even the image of the wild, formally restoring highly managed populations of large carnivores through efforts such as the Yellowstone Wolf Project.

This was also a period of time prior to which there were only rudimentary forms of the Internet and cell phones, so there was no spontaneous way to contact the outside world or be contacted. No global positioning units to provide directions; only paper maps, a compass, and memory. It was a time when you still could get lost in relatively unexplored corners of the West, imagining that if you veered just a little off the trail you would step on a patch of dirt where no human had previously placed a meandering foot. The view around the corner on a hiking trail was visible to you only as a reward for your efforts, and not to millions of people simultaneously through real-time satellite imagery, Google Earth, or a strategically placed solar-powered webcam.

Now, after spending a large portion of my life traveling in remote parts of the Great Plains, when I find a prairie dog town without black-footed ferrets it is hard not to think about what used to be or perhaps could be again. Similar to the finest pieces of art in a museum, to me, ferrets are that rare piece of the ecosystem puzzle that not only makes the prairie more noteworthy, but more complex and beautiful. Not only because ferrets are a unique trademark of the plains, but also because they represent a long history of struggle to save them from extinction and restore them to their prairie home. It is this past that makes any patch of land a ferret now occupies a little more memorable and special, for that land has regained a sense of being more complete and wild, a wildness that existed before, and can exist now only with gentle human intervention.

In this way, black-footed ferrets represent the wild heart of the Great Plains in an increasingly modern and civilized age. The question that remains unanswered is whether people will tolerate ferrets and their prey and allow them to recover—whether society increasingly finds value in reviving and rewilding the Great Plains.

CHAPTER I

Pleistocene to Anthropocene

The true West is defined not by time zones, but by geology, soil, and water. By grasses and sedges, wildlife and openness. Driving west across South Dakota on Interstate 90, as I cross the Missouri River at Chamberlain, the land changes from flat agricultural fields to rolling native prairies. Irrigated, domesticated green gives way to cattle pasture that remains a natural brown on pitches and breaks too steep to plow. I breathe deeply to take in the smell of grasslands. Despite spending an entire day in them driving up from Kansas City, I finally feel the sense of entering into the Great Plains.

This is a land of unfolding views, where days are not numbered in the mind but remembered for the weather, the clarity of the sky, the heat of the sun, and the strength of the wind. Today was a sixty-mile-view day limited only by the curvature of the Earth. I pass the interstate towns of Murdo, Kadoka, Cactus Flats, and Wall, with gas station and hotel economies sustained by the needs of travelers. Other small towns just out of view of the interstate or without exit ramps are abandoned and desiccated.

At seventy miles an hour, interstate thoughts are quick. They pass faster than mile markers and are as easily forgotten. Traveling at this speed, everything looks similar, flat, like the parallel lines on the road. But if you take time to slow down, you see that the plains are far from static and monotone, and if you stop for a while, the land changes. Live here for a year and you will appreciate the seasonal cycles of winter and

FIGURE 1. American avocet *(Recurvirostra americana)* on the edge of a prairie pothole wetland in South Dakota.

summer and the temperamental periods in between. The dryness and wind that dominate the land keep vegetation low to the ground and flexible. It is ready to sprout with a spring rain, early to reproduce, and then quickly senesce to a gentle brown for the next nine months of the year.

Where there is water on the plains, there is intense life and color. Find a pond in midsummer and you will have a pair of marbled godwits swoop down on you with four-inch bills and eight-inch legs, defending their nest. Hear a hummingbird-sized marsh wren singing, more melodious than the nightingale and more wild, with white eyebrows and upturned tail. See black-necked stilts sparring with American avocets on the marsh edge. Hear the whimpering of a Wilson's phalarope as you move past her nest. See a pair of competitive male yellow-headed blackbirds perching on the small yellow blooms of sweet clover. In a nearby cottonwood tree, northern orioles and yellow warblers will flash their colors as they hang and lunge for insects. There may be house wrens nesting in a rotted-out knot of a tree branch and flickers and downy woodpeckers poking and prodding in its pale, deeply creviced bark. In the willows, evening grosbeaks and yellow-breasted chats may be moving and calling to whoever will listen.

Away from the rare river, stream, or pond, the land of the Great Plains is best described by soil and distance from the Rocky Mountains. The rain shadow of the Rockies causes waves of prairie ecotypes: shortgrass, mixed-grass, and tallgrass. Farthest west, the sagebrush-dominated high plains of shortgrass prairie receive as little as ten inches of rain a year, too dry for trees but ideal for grasses that pierce the earth to hold down soil with minute shallow root hairs that can grow sixty miles a day. An irrigated hay field can exhale five hundred tons of water a day, but the economical native grasses of the prairie are adapted to the cycle of water conservation, timing their growth and reproduction to the seasons.

I follow the interstate west over the gradual rise of the Black Hills and enter Wyoming. Passing through the gas and coal mining city of Gillette, where a boom in natural gas production has caused a Grapes-of-Wrath style migration of workers from small towns in the Midwest and beyond. Streets are crowded with families who drive new trucks and SUVs but live in shared trailer homes—temporary homes for people who are flush with money from the energy fields but have to wait for construction to catch up with the influx of immigrants. Outside of town, new roads branch out to gas wells in all directions, harshly scraped by bulldozers, looking like varicose veins on the sagebrush steppe.

Continuing west, the snow-capped Big Horn Mountains force me to shoot north to Sheridan. By late afternoon I am across the Montana border into Billings, and get off the interstate to take the smaller route US 87. Now heading due north, I pass through the rolling hills north of Roundup, where my friend Mark is stationed as a state trooper. With so few cars and little to do, he would often pull me over when he recognized my truck, so that he could talk of his latest exploit. Last time it was how the week before he finally got to shoot his sidearm in order to finish off a porcupine, half run over and languishing in the middle of the highway.

North of Roundup, the prairie opens up for a 157-mile stretch, with the next town being Malta, followed shortly thereafter by the Canadian border. The sun is setting, turning the sky to a fading glow that puts shadows on grasses by the edge of the road. Nighthawks come out on their saber wings to pick off insects that are stirred up by the cooling temperatures and waning light along the roadside ditches. Just after dark, I dip down to the Missouri River, cross the Fred Robinson Bridge, and climb up the hill on the other side. Now within the million-acre Charles M. Russell National Wildlife Refuge, I take the first dirt road

FIGURE 2. Mixed-grass prairie homestead in central Montana.

east, driving until I know the highway is out of view, and pull over to spend the night in the cab of my pickup.

. . .

I awake to cool morning air and a flaming orange and pink sunrise. Fingers cold, but not quite numb. I slept in my faded blue jeans. I trade my hiking boots with bent grommets and knobby Vibram soles for flat-bottomed cowboy boots that are worn to fit, slip on easily, and won't cake with mud. No traction needed here. A killdeer cries. Sparrow-sized horned larks battle, black tails flapping.

When I step out of the truck, a mixture of western wheatgrass and blue grama crunches under my feet. They are just tufts of dried brown grasses on the surface, but below ground they form a maze of root hairs that teems with mycorrhizal fungi to access nutrients, tap into moisture, and store energy below ground. Root systems have been built and added to over generations so that 75 percent of the prairie's biomass resides below ground, roots that are lost when the prairie is plowed and turned over for row-crop agriculture, the land losing three-quarters of its living biomass in one fell swoop.

Scientists have shown that it is not tropical rainforests or arctic tundra but rather temperate grasslands that are among the most quickly declining ecosystems on the planet. In North America, grasslands have the inglorious label of "the most endangered ecosystem." It feels good to be in the middle of one of the largest intact stretches of mixed-grass prairie left in the world.

I once chose these grasslands over the woman I loved. It wasn't a sudden choice, rather an accumulation of days, weeks, seasons, and years that taught me to get in my car and head to the prairie. A slow burn that led me to value a familiar place above all else.

Finding a neat line of rocks with multicolored red and green lichens that form a circle, I recognize a teepee ring. I feel satisfied that my campsite selection was appreciated by those with deeper bloodlines and far greater knowledge of the area than mine. Flat skylines extending in all directions show rolling treeless prairie, with a small patch of hills to the east known as the Little Rocky Mountains, a fine vantage point from which the Gros Ventre could watch out for raiding Assiniboine and Cree tribes. Perhaps it was near a bison migration route, so they could watch and wait for the seasonal movement of the herds from southern Saskatchewan south to the center of Montana, an annual cyclical movement pattern back and forth between summer and winter that had become entrenched, dictated by topography, water, and forage. Following the same path year after year, herds moved north from more mild wintering grounds in perfect timing with the spring green-up of the Saskatchewan prairies.

I think of the infinite magnitudes of bison described in the journals of Lewis and Clark and later painted by George Catlin in the 1830s, a time when 25–40 million bison roamed the plains and were the mainstay of regional native tribes, before disease, firearms, and uncontrolled hunting reduced the giants to fewer than a thousand individuals by the 1880s. I had last seen one of those remnant animals in Yellowstone National Park, walking in the open valleys between forested peaks, far from the center of their vast Great Plains home. The last herds were always in fragments, lost from the core.

Now the grasslands seem silent and empty without them. When Lewis and Clark traveled along the upper Missouri, they encountered the now-extinct plains grizzly and Audubon's sheep. Wolves still coursed over the prairie chasing elk and bison. Whispers in conservation circles of bringing bison back here evoke feelings of guilt, mixed with hope. Guilt, because of what we have done and that we still need to only whisper. Hope, because beyond regret, there is a selfish and at the same time selfless sense of purpose in righting a wrong accomplished through greed and mistaken assumptions of infinite magnitudes. A sentimental common thread that has continually driven many conservationists to action in the past, and will continue to do so in the future. I think of the group of extremist ecologists who want to go further: to replace extinct North American mammoths, *Camelops* camels, and Haberman horses that were unable to withstand the Holocene megafauna extinction period caused by environmental change and human hunting. As replacements, they call for introduction of African elephants, dromedary camels, plains zebra, and other surrogate species in a process called "Pleistocene rewilding." By doing so, they would dramatically shift the timeline for conservation from two hundred to ten thousand years ago, making current efforts to reintroduce bison and wolves seem mundane compared to their proposals for elephants and lions.

To find justification for their mad plan, one has to look no further than the pronghorn antelope, the swift trademark of the western plains with speed that today seems an evolutionary misfire in excess. But nature is never overly generous without necessity, and it is easy to envision how now-extinct American cheetahs forced the pronghorn to such speeds. After a lifetime of nature documentary scenes of African cheetahs chasing Thomson's gazelles across the Serengeti, I quiver at the thought of seeing a similar scene play out with pronghorn on the Great Plains.

. . .

The sun finally breaches the horizon, sending out stark white light that bleaches away the oranges of morning. I walk down to a nearby water tank to wash my face. There is the familiar smell of fresh cow manure and exposed dirt. The trampled ground is sparsely covered with grass and sage tufts, grazed by cattle to the smoothness of an old river bottom. Surrounding the tank, brown fluffy prairie dogs sit hunched over, balls of fur warming in the morning light. Clouds the color of gray sagebrush drift overhead. Cold clumsy mosquitoes fly into my hair.

A prairie dog cries. Mother to son? Son to sister? Family. Do others listen as well: the badger, the coyote, or the owl? Owls typically hunt at night, but I have seen a great horned owl leave its day perch in a Russian olive tree to swoop down on an unsuspecting prairie dog, squeezing tight and puncturing with sharp talons, forcing wheezy cries and a stream of urine from its prey, lingering on the ground for a few seconds before struggling to fly back to its perch with the two-pound meal.

It is not by chance that prairie dogs are here. They seem to hate vegetation more than an inch tall, or anything else that could obstruct their view of an approaching predator. To the distress of golf course managers and cattle ranchers, they thrive on land where intensive mowing or grazing has reduced grasses almost to bare dirt. The close ancestral relationship between prairie dogs and bison as grazing partners likely created a mosaic of prairie dog colonies across the landscape linked to the migratory pathways of bison. Colonies of hundreds of thousands of individual prairie dogs once overlapped those migratory pathways, permanently occupying both the summer and winter ranges while their larger partners moved along.

When Lewis and Clark ventured onto the plains from the Missouri River that flows just below me out of view, they often took note of the guinea-pig-sized mammal. At one point stopping long enough to experiment:

> CLARK—FRIDAY, SEPTEMBER 7, 1804
>
> Discovered a Village of Small animals that burrow in the grown (those animals are Called by the french Petite Chien) Killed one and Caught one a live by poreing a great quantity of Water in his hole we attempted to dig to the beds of one of those animals, after digging 6 feet, found by running a pole down that we were not half way to his Lodge. . . . The Village of those animals Covd. about 4 acres [1.6 hectares] of Ground on a gradual decent of a hill and Contains great numbers of holes on the top of which those little animals Set erect, make a Whistleing noise and whin allarmed Step into their hole. we por'd into one of those holes 5 barrels of Water without filling it. Those Animals are about the Size of a Small Squ[ir]rel . . . except the ears which is Shorter, his tail like a ground squirel which they shake & whistle when allarmd. the toe nails long, they have fine fur.

Prairie dogs, along with magpies, were among the only living things Lewis shipped back to President Jefferson in 1805 from Fort Mandan, where the expedition stopped for its first winter. Despite the long trip, prairie dogs were robust enough to survive from North Dakota to Jefferson's desk at Monticello. Jefferson kept one prairie dog as a pet for a time before passing it on to Charles Wilson Peale's museum in Philadelphia, where it lived out its days as an attraction symbolizing the novel and unknown American West, then was stuffed as a curiosity for decades more, and finally was lost in a fire.

. . .

For all their interest in prairie dogs and other wildlife, Lewis and Clark never observed a black-footed ferret. Indeed few European explorers had seen them other than as pelts used by indigenous people. Spanish explorer Don Juan de Oñate was perhaps the first European to describe the species in 1599 while exploring the future southwestern United States: "It is a land abounding in flesh of buffalo, goats with hideous horns, and turkeys; and in Mohoce [center of the Hopi nation near present day Walupi, Arizona] there is game of all kinds. There are many wild and ferocious beasts, lions, bears, wolves, tigers, penicas, ferrets, porcupines, and other animals, whose hides they tan and use."

Although Oñate could have been describing bridled weasels or other weasel species known to exist in the region, ferrets were likely present in the area. People of the Blackfoot, Cheyenne, Crow, Hidatsa, Mandan, Navajo, Pawnee, and Sioux nations have all used black-footed ferret hides in the making of skins for headdresses, medicine pouches, tobacco pouches, or other sacred tribal objects. They also have had distinct names for ferrets that illustrate their familiarity with the species and its biology. The Sioux called ferrets *pispiza etopta sapa,* translated to "black-faced prairie dog," illustrating their knowledge of the key link between ferrets and prairie dogs. The Pawnee called ferrets "ground dogs" in one of their mythical stories where the ferret speaks of itself as "staying hid all the time," which shows the Pawnee's familiarity with the reclusive nature of ferrets.

Fur trappers during the early 1800s also were familiar with ferrets and differentiated the species from other mustelids (the family of mammals containing stoats, mink, wolverines, otters, and other weasel-like elongated carnivores) before scientific discovery and classification. Pratte, Chouteau and Company of St. Louis, better known as the French Fur Company (and later as the Western Department of the American Fur Company) concentrated their fur acquisition efforts in the "Sioux country" of the upper Missouri River basin encompassing most of present-day Montana, Wyoming, and South Dakota. They listed eighty-six black-footed ferret pelts received between 1835 and 1839, a taxonomic distinction not yet known to science, but that the trappers noted apart from "weasels" on their ledger.

It was trapper Alexander Culberson who first brought black-footed ferrets to the attention of the eminent naturalist John James Audubon. John Bachman and Audubon provided the first scientific description of the species in 1851, based on a specimen collected near Fort Laramie, Wyoming. Unfortunately, this original specimen was lost, and the validity of Audubon's discovery was questioned by naturalists for the next twenty-five years. Even with the pedigree of Audubon supporting its existence as a species, the validity of the reclusive ferret of the Great

Plains remained a topic of debate until 1877, when Smithsonian curator Elliot Coues was able to procure several additional specimens to confirm Audubon and Bachman's classification.

This debate persisted despite the fact that prairie dogs, on which black-footed ferrets rely, were likely to have been one of the most abundant mammals in North America at the time. Naturalist C. Hart Merriam noted the abundance of prairie dogs when he made transcontinental train journeys across the Great Plains in the late 1800s. He wrote that "the traveler who looks out the car window by the second day west from Chicago is sure to have his attention arrested by colonies of small animals about the size of cottontail rabbits." Merriam noticed that they had become tame to the sounds and sight of the onrushing train, staying above ground long enough for him to take notes on their colonial life and make observations on their social behavior. He reported on their elaborate burrow structure, their seasonal cycle of activities, their warning calls to others, and the species that preyed on them. Merriam reported hearing of a colony in Texas estimated to be twenty-five thousand square miles in size, and based on the density of burrow openings, he deduced that it contained 400 million prairie dogs.

At a continental scale, he postulated that prairie dogs had an inflated abundance "due to the coming of the white man" who "cultivates the soil and thus enables it to support a larger number of animals than formerly." Although intensive grazing of the grasslands by overstocked exotic cattle created a more open prairie that might have allowed prairie dogs to increase in numbers, the species was always abundant across the Great Plains, even prior to westward invasion by the "white man." Prairie dogs had been in the Great Plains for millions of years prior to Lewis and Clark's expedition, with fossil records dating back to the late Pliocene epoch of 2.5–1.8 million years ago.

It was the high abundance of these uniquely New World rodents that allowed for the evolution and speciation of one of the world's most specialized carnivores, the black-footed ferret. Similar to bison and many other New World mammals, black-footed ferrets can trace their predecessors back to Europe and Asia. The precursor to the modern black-footed ferret, a now-extinct subspecies of the steppe polecat *(Mustela eversmanii berengii)*, followed the Bering land bridge from Asia to North America. Slowly spreading southeast from modern-day Alaska through ice-free corridors, this ancestral ferret was present in the Great Plains as early as eight hundred thousand years ago.

From the specialized beaks of Darwin's finches to the cryptic partnerships between flowers and their uniquely dependent pollinators, ecologists have always loved to study specialization—the ability of species to adapt and form a dependence on a specific set of conditions. Over time, this new set of specialized adaptations can become so advanced that it forces a species to diverge, become unique, eventually forming a separate species. In this way, specialization itself is the foundation of biodiversity, but to be able to specialize requires three things. First, there must be an advantage to specializing, to doing one thing better than any close relative. Second, the selected advantage must be heritable and able to be passed down through generations. Finally, there must be stability in the advantage so that the plusses and minuses of reproduction and survival favor the specialist over its competition.

The prairie dog populations of the Great Plains provided the perfect medium for specialization of the black-footed ferret. They offered an abundant and stable prey source that allowed some ferrets to begin to shift their diet from a range of small rodents similar to those found in Asia and Europe to the larger prairie dogs of North America. Further, the intricate burrow systems of prairie dogs served as sufficient shelter for these prairie-dog-hunting specialists. With both food and shelter available, these early black-footed ferrets found no need to leave prairie dog colonies and interact with their ancestral predecessors, and by as early as thirty-five thousand years ago, the black-footed ferret *(Mustela nigripes)* was morphologically distinguishable from *Mustela eversmanii berengii.* Speciation had occurred, producing the highly adapted black-footed ferret that has been able to persist long after its generalist polecat predecessor had gone extinct in North America.

. . .

As first chief of the Biological Survey (later to become the present-day U.S. Fish and Wildlife Service), Merriam focused his interest on patterns of species distributions in North America, which he mapped as "life zones" or biomes. On first observing how prairie dogs and their distribution coincided with prairie grasses but avoided belts of trees in stream valleys, he found support for his theory, musing that prairie dogs were "an important illustration of the law that in fixing the limits of distribution of animals climatic factors are even more potent than food."

Merriam also had thoughts as an early conservationist, even contradicting the original intent of the Biological Survey, which was to provide information on birds and mammals that were agricultural pests. In his 1896 report to Congress, he opposed the "pernicious effects of laws providing bounties for the destruction of mammals and birds." He did not, however, extend these sentiments to prairie dogs. His 1902 report entitled "The Prairie Dog of the Great Plains" reflected public pressure from the rapidly growing number of western farmers and ranchers. In this seventeen-page report, he institutionalized a view that has largely persisted since: that prairie dogs are a rangeland "scourge" that must be eradicated.

By the late 1800s farmers and ranchers had moved into the plains in such numbers that land holdings were decreasing in size, and grass losses due to prairie dog herbivory were "more keenly felt." Merriam postulated that prairie dogs resulted in the loss of 50–75 percent of the production capacity of a piece of land for livestock (a figure widely discredited in the years to come). This assessment, combined with the view that prairie dogs were abnormally expanding their range following western settlement, led western states to develop policies and programs for the complete elimination of prairie dogs. Texas and Kansas even went so far as to adopt legislation around 1900 that stipulated fines for landowners who failed to exterminate prairie dogs on their land.

Because of the low price of strychnine and the financial repercussions of inaction, prairie dogs were removed from large portions of Texas, Kansas, and other states throughout the West. Even after millions of acres of prairie dogs on private farms and ranches had been eradicated, there were still complaints about prairie dogs on government lands where private landowners leased grazing rights. Merriam referred to this situation as a "very serious evil" for private landowners, "and one with which it is exceedingly difficult to cope."

The federal government, with what Merriam termed "universal support," started to investigate efficient ways to "combat the evil" on private and public lands. Prairie dog eradication changed from a backyard occupation to a federally funded campaign. The government purchased and applied poison grain laced with strychnine and provided subsidies for ranchers to do so on a colossal scale. More recently, strychnine was replaced by 1080, zinc phosphide, and Rozall, but the effect was just as lethal. After treatment, sites were revisited to ensure a complete kill. Stragglers were knocked out with targeted burrow application of bisulfide, pyrotechnic gas cartridges, cyanide flakes, or a homemade concoction called "hoky poky on cobs."

Prairie dog poisoning became a full-time job for thousands of people. Eradication was a sure-fire campaign platform for any governor or congressman in a prairie state. Poisoning was both politically popular and well funded in order to remove an "impediment to the economic development of the west." Between 1915 and 1965, western states eliminated more than 37 million acres of prairie dog towns, driving populations to less than 5 percent of their former range. They left only small, overlooked pockets of prairie dogs in the back corners of a few expansive ranches, public grasslands, and tribal lands. These pockets today continue to be targeted by government-sponsored poisoning campaigns, with only some small populations being managed in protected environments.

. . .

I return to my truck and leave the prairie dogs to start their day. I find in the bottom of my backpack a granola bar, oats and other grains molded into a snack that is thankfully strychnine free, and think of C. Hart Merriam as I eat my breakfast. Reading his 1902 article, I thought that Merriam sounded interested in the ecology of prairie dogs. He seemed almost fond of prairie dogs when he referred to the black-footed ferret as "one of their [prairie dogs] most relentless and terrible enemies, and if sufficiently abundant would quickly exterminate all the inhabitants of the largest colonies." Yet his document now can be seen as a masterful piece of political propaganda aimed toward securing funding for his agency through prairie dog eradication.

I wonder if he ever realized, or possibly regretted, his role in forever shifting public opinion of prairie dogs as a symbol of the West to that of a pest. Even the founder of modern wildlife management, Aldo Leopold, originally subscribed to the eradication view for a time, shooting wolves on sight in the Gila National Forest of New Mexico where he was a forest ranger. Yet one day while staring down at a wolf he shot and seeing the "green fire" extinguishing from its eyes, he had a vision for the future that would forever change the way we view and manage wildlife. Leopold's feelings of sentimentality and mourning brought about the birth of a land ethic and greatly advanced the conservation movement. Unfortunately, C. Hart Merriam did not have a similar vision for wildlife. History now remembers him as a man after whom several wildlife species were named, many now extinct, with the most notable exception being a subspecies of turkey.

Today, natural selection seems to favor the generalist: the raccoon or dandelion that persists in the city, suburbs, and national park; the grub that can feed on multiple backyard plants rather than be restricted to one, thereby minimizing its vulnerability to the plight of a single bush. As the human footprint expands and homogenizes the landscape, specialists are often those rare holdovers we both marvel at and want to conserve—species like the panda of the bamboo forests in Asia, desert pupfish of small pools in the deserts of the southwestern United States, and the oyster mussel of clear Appalachian streams, that are at the top of many conservation priority lists as the species at greatest risk of extinction.

Just as ecological theory would predict, the decline of prairie dogs has been accompanied by the decline of their specialist predator. Black-footed ferrets have always been cryptic and difficult to find, but by the mid-1900s they were becoming exceedingly rare. In the span of less than a hundred years, the black-footed ferret would go from the high status of being the most effective specialist in North America to one of the most critically endangered animals in the world. By the late 1960s and early 1970s ferrets would become known not as the "most relentless and terrible enemy" of the prairie dog, but—thanks to legal protection afforded them under the Endangered Species Act—as the saving grace for their prey species. Endangered predators relying on declining prey, the ferrets became the bellwether of healthy, large prairie dog colonies on the handful of sites where they still existed.

I look out the truck windshield past the prairie dog colony I had just left. In the distance I see a series of other small colonies on benches leading down toward the Missouri River. They are just patches of bare ground and burrows on the flat tops of hills separated by narrow draws, where I know that at least a handful of prairie dogs would be coming above ground for the day. Through government protection and proactive restoration, prairie dogs have persisted here over time. It is a small victory in the long-term battle to save the few last pieces of intact native prairie. I hope that someday, in my lifetime, we might be able to restore ferrets back here. To have the rarest eat the rare. It would take me hours to get anywhere with a name on the map, but I am in the middle of everything.

CHAPTER 2

Decline toward Extinction

During summers when I was young boy, I would go with my father to his office. In the early morning, the building was filled with scientists crossing paths as they either headed into the field or started their day of typing on computers. Barely knowing how to type, I volunteered to work in the field with Rob Hinz. Rob was a pastor at a local church on the weekends and trapped meadow voles for scientist James Nichols during the week. Fifty years old, with a round belly and balding hair, Rob always wore three layers of heavy cotton shirts, even on the hottest of days.

"Sweat keeps you cool," Rob would say. Water kept against skin. I found no reason to doubt him; his logic seemed sound.

We began checking traps at dawn so that we would be done before the heat of the day, working our way up and down a grid of trap lines in the tall, humid fields of Maryland. We checked to see whether the door was shut on each small, shiny, aluminum box trap that was placed under a neatly cut, rectangular square of plywood to keep it from baking in the sun. Each trap location was marked by a four-foot wooden post planted into the ground, painted white on the top so we could find it in the thick grass.

As we walked through the grid, checking and closing traps, Rob would chain-smoke cigarettes. Finishing one, Rob would kill the ember on the top of a white post, pinch off the quarter-inch of tobacco from the filter and neatly leave the white tuft on top of the post. "Something for the deer" he would say as he placed the used filter in his canvas shirt pocket before pulling out another stick. He wheezed as he walked in the heat, fifty feet between traps, pouring sweat from his brow that dripped down into the folds of his neck and was absorbed in his faded cotton layers. Yet when he had an animal in hand, his round plump fingers were dexterous and gentle. He held the small brown vole barehanded, gripping it by the scruff of its neck, knowing the intent of its sharp incisor teeth but avoiding little nips, reading me data to record on the clipboard. Trap 5M, female, VP, medium nipples . . . 230 grams.

Finishing our grids for the day and resetting traps, we drove back to the office, passing ponds and mowed fields that served as study plots for other scientists. Groves of mixed hardwood trees covered with little metal tags and brightly colored flagging, signs of scientists at work. On the forest floor, metal cages and trays lay at set intervals, collecting falling leaf litter. I wondered who had that job, who found something of interest in what I thought was ordinary. Taking the route past ponds filled with cricket frogs and painted turtles, just before the gravel road gave way to pavement, we passed by a dirt road between two ponds that was blocked off with a sawhorse and a sign that said "Restricted Area—Whooping Crane Staff Only." On quiet days in the surrounding fields, I could hear the sounds of cranes coming from across the motes of ponds. Goose-like calls that were more awkward and guttural. Having seen pictures in my bird field guide, I envisioned the vibrations starting far down the cranes' long white necks, emerging from their beaks opened slightly ajar and serving like the flare of a trumpet. I could only imagine what existed beyond that sign. Compared to the common field mice and frogs I was used to, the cranes were mysterious, critically endangered dog-sized birds from the Great Plains. They were a first taste of the sentimental and at the same time selfless exhilaration of rarity, knowing something so rare was being carefully hand-raised in Patuxent, Maryland—a protected oasis surrounded by freeways, city parks, and subdivisions.

. . .

In the afternoons I would sit with my father in his office, eating my lunch and watching him eat without stopping work. A field biologist turned administrator, he would awkwardly hunt and peck for keys on a computer while slowly eating Fig Newtons and an apple. It was the same small lunch every day, washed down with four or five cups of coffee. Being otherwise surrounded by field biologists, I couldn't yet understand how he found contentment in a thirteen-hour-a-day indoor work routine that included trips in on weekends to "catch up."

It was only that summer that I learned that to a former field biologist, "catching up" meant getting away from the computer and phone he was tied to during the workweek. Because of a recent change in leadership at the research center, the historic Merriam Laboratory was being put through a cleaning frenzy. As with any political regime change, the new director wanted to make a gesture of renewal. Out with the old and in with the new by way of dumping an attic full of old note cards and field notes to make room for the current and next generation of biologists who needed room for their old laptop computers and floppy disks. On Saturdays my father would make the one-hour drive to Patuxent from our home to comb through the old file boxes stacked by the Dumpster, beating the Monday morning trash pickup and squirreling away boxes of unpublished, original reports from Aldo Leopold, letters from C. Hart Merriam, and other irreplaceable historical documents from an era before compact discs and hard drives.

"Last weekend I found a box of ferret pictures you might like to see," he said, knowing I had been pestering him lately about getting a pet ferret, a European polecat from the local pet store just like my neighbor, Brian Peel, had.

"Wild ferrets?"

"Yes, very rare black-footed ferrets from out West. Much neater than the ones at the pet store."

My only experience with the West as a child was books and movies and the Zuni fetishes my mother had purchased for me on winter trips to Arizona with her sister. On her latest trip she bought a book for me telling of the ceremonial symbolism of the fetishes. It was a form of religion I could wrap my head around, animals at its center, based upon the land.

At night, I would set up the Zuni fetishes in my attic bedroom—small carved rock animals that I kept packed in a woven pine needle bowl with a tight-fitting lid. I wrapped the smooth, polished, brightly colored stone animals in scraps of leather I found in the basement and bits of sage mailed from my uncle's ranch in Wyoming. I would gently unwrap the fetishes, remembering the species and stories but also creating my own as I lined them up on a table by the window. Hearing the slowly lapping waves on the Chesapeake Bay through the old oak trees outside my window, I would light candles, turn out the lights, and feel the individual fetishes in my hand as the book said, squeezing tight and letting go, closing my eyes and trying to let go, hoping for a vision.

With no luck, I would repeat the ceremony on another night, arranging the figurines in a different order, leaving some in the bowl. I thought it might be my trappings that blocked the spirits. I tried burning some of the sage to clear the air as I had seen in movies. I took my shirt off to clear a pathway to my heart, where I was told my soul resided. Seconds felt like minutes that felt like hours as I waited, hurriedly blowing out candles and putting my shirt on only as I heard my father coming up the stairs.

The visions never came.

After I finished my sandwich I went up to the attic of Nelson Lab. It was the only room without the bustle of scientists typing on keyboards. My father gave me the padlock combination, and fixing the numbers in place, I released the white-painted door that stuck tight from the humidity and lack of use. Inside, the attic was dark, wooden, stuffy with the smell of old paper and dry, rotted leather. A thirty-foot row of wooden file card cabinets filled the middle of the room and contained five hundred thousand index cards with bird observations dating back more than a hundred years. Metal cabinets filled with measuring tapes, binoculars, and metal tree tags lined the outside walls and made small alcoves. Old cardboard boxes were stacked in corners, filled with field notebooks and research reports from projects finished or abandoned—questions were answered, funding ran out, forests were cut down, researchers moved on, species went extinct.

I walked a lap around the island of wooden bird-card file cabinets and found the stack of cardboard boxes he had set to the side, marked "Do Not Discard." Inside the first box, I found a stack of papers with original large glossy black-and-white photographs of a black-footed ferret. There were also letters, memos, and reports from Montana, South Dakota, North Dakota, Wyoming, Saskatchewan, Colorado, New Mexico, Nebraska, Utah, and Kansas. The oldest was a typed letter by a South Dakota trapper, Ralph Block, who reported his first ferret sighting in 1947: "The first observance I ever had of one was in northeast Nebraska, Knox County. I trapped this animal with a steel trap and wondered what this strange looking animal was and thought this was a freak mink. I had not entirely concluded this idea, but sent it in with a shipment of furs to a fur house, Sears Roebuck. In a few days my check for the other furs came along with a notation on the grading chart, noting that there was also a pelt of no value."

. . .

Each day, after a morning of trapping voles, I would return to the attic and stay late into the afternoon while my father worked in the office below. I read through dozens of letters from farmers, ranchers, housewives, and professors. All of them reported to have seen a ferret on their property or crossing a road at some point during the 1950s and '60s. Most were obviously inaccurate, and all were nearly impossible to verify or highly questionable, like this letter from John Francis:

> About 6 years ago in August or September I took a trip up to near Lemmon S Dakota to drive a tractor. I worked this plow and tractor work 3 seasons. On one trip I visited a coffee buddy's cousin that had migrated to there from Lewis, KS. He had a real type sheep ranch, very well kept with clean water by well and pump and beautiful fields of alfalfa and grass and wheat. I drove my car to his sheep ranch and visited. I had an auto track for about a mile with thick weeds bumper high. I was watching my trail in, when a possible ferret or weasel of kind jumped about ten feet ahead of me and cleared both trails in a 2-second of time and into the thick high grass. I could only see that he looked like a long stove pipe, and was used to a car or pickup, and was out of sight.

The letters by trapper Ralph Block were consistently the most reputable and interesting. I pictured a short, stocky man who would always be found in plaid shirt and cowboy boots, sporting a mustache and driving an old pickup truck covered with a film of dust from prairie dirt roads. In 1949, Ralph was hired by the U.S. Fish and Wildlife Service for predator and rodent control, stationed out of Isabel, South Dakota. He began to spend extended amounts of time on prairie dog colonies as he carried out his job to exterminate them. He submitted typed stories of his interactions with black-footed ferrets, making repeated trips to see them during his trips to "clean up" remaining prairie dogs after a poisoning event and occasionally on his weekends and free time. A cowboy conservation biologist, he was killing the West while also beautifully recording its biota for posterity. After a week of reading, I concluded that Ralph had come up with more sightings than all of the others whose reports were in the file.

> In the spring of 1950 my second observation of a black-footed ferret was at a location about 14 miles southwest of Isabel on what is called Corn Creek in a large prairie dog town, on the Forrest Thompson Ranch, Section 24 T. 15 R.21. This was an adult male which I procured with .22 rifle, froze, and packed in dry ice and now on exhibit in the National Museum at Washington, D.C. This was an adult male estimated (I believe if I am remembering right) 4 years old, was about 22–24 inches long, carrying several wood ticks plus some gopher ticks. I received a very complimentary letter regarding this from the institution as they had never been able to get a single specimen and at the time it consisted of 4 full floors of exhibit.
>
> In the summer of 1950 the service entered into an agreement with the Rosebud Sioux tribal enterprise for control of extensive prairie dog towns on the Rosebud Indian Reservation. This was a project that involved the use of 1080 poisoned oats (Sodium Fluoracetate). This program involved the use of

a crew of native Indians who assisted me in treating large holdings of prairie dogs on range lands at the request of landowner, lessee or tenant with possession of the land (with signed agreement-waiver release). I made my headquarters in Rosebud, had a crew of from 5 to 6 men, and during that summer treated well over 34,000 acres. It was during this summer that I saw two ferrets living in two separate prairie dog towns. At this time I started getting interested in them myself. One was living at a place between Norris and Belvidere, South Dakota, on the Baxter Berry Ranch, of the late ex-governor Tom Berry. There was a terrific acreage in this dog town (several sections). In fact, the village of Norris was completely surrounded on all sides by prairie dog town. Grass was almost non-existent and even the grass roots in the sod were eaten by these rodents. Land owners paid for the grain at cost and wages of the crew on a revolving basis.

The next ferret was this same summer in a prairie dog town on a fork of the White River northeast of Blackpipe day school on Indian land operated by Indian ranchers. I believe their name was Blue Dog (I have since lost notes on this).

It was during this summer that I replaced Walt Stammerjohn, MCA from western Nebraska loaned to South Dakota for assistance as our group of hunters numbered only 4 and the agreement stipulated a full season of rodent control on the reservation.

On ferret #4, the last mentioned, the prairie dog town was also treated with 1080. Since there was an early day fort located in the area on the banks of the White River tributary, I visited this area on weekends and every chance while nearby would drive over to the particular area when I first noticed #4. In nearly every instance there he would be, nearly like waiting for me. I would just stop in the area, wait a while and he would emerge from a prairie dog hole. Sometimes just part way out, other times entirely out. I especially want to mention that this town was a 100% kill of prairie dogs.

The following summer [1951] an extensive program was carried out on the Pine Ridge Reservation. . . . I used 3 crews of Indian labor putting out something just over 12 tons of 1080 on 43,137 acres. It was this summer also that 2 different ferrets were seen and both of these "survived" the treatment. One near the village of Kyle and the other in the vicinity of Oglala. John Limehan was owner and operator; also he operated a small country store. At that time there was an Indian agent or farm agent, Willis Adams, who assisted me a lot, arranging for lunches, helping to hunt up owners (mostly Indian operators) for releases, inquiring of finances and the like. He was one that I took along who got a look at the last mentioned ferret (#6). The Superintendent (Sanderson) at that time of Pine Ridge Agency once went along to supposedly also look at it, but, of course, wandered off in a nearby untreated section with his .22 rifle obviously unmoved to anything but the excitement of shooting off a few prairie dogs.

. . .

The first attempt to restore black-footed ferrets was linked to Walt Disney of all people, in his company's production of the film *The*

Vanishing Prairie. In 1953, federal predator and rodent control agent George Barnes was enlisted to live trap three black-footed ferrets from Fall River and Jackson Counties, South Dakota, and release the ferrets into Wind Cave National Park for the film crew. Of the three he brought to Wind Cave, only one survived until January 1954, but the precedent was set. The survival of at least one suggested restoration was possible, and the park requested six more ferret pairs. To increase the supply of ferrets for relocation, the State of South Dakota mailed a specially designed round aluminum trap to each rodent control specialist working in the western part of the state. With Disney-like visions of heroes and villains, my young mind could not yet wrap itself around the thought of enlisting the prairie dog exterminators to save the declining predator. I couldn't understand how biologists who were hired to kill, wholesale and without remorse, the singular prey species on which ferrets depend, were at the same time trying to save the predator.

Over the following years, Ralph saw ferrets on multiple prairie dog towns across South Dakota, including one to the north in Haakon County near the Cheyenne River:

> Not over a mile or mile and a half from the Bridges Day school on land owned by Chicago Cattle Co. (now known as Western Cattle Co.)—owner Wm. Norton. This one the crew also saw; in fact, it sat or stood immobile all the while we ate lunch nearby—almost like a statue. By the way, they can be caught easily. At that time the area was being extensively tested for oil leases. They used long electric detonator wires of plastic coating wound loosely. They made ideal snares.
>
> In the summer of 1956 I worked a joint program on Standing Rock Reservation, Sioux County, North Dakota, Carson County, South Dakota. While working I stayed at Fort Yates, North Dakota. It was then that one more ferret was observed in a prairie dog town in just about the exact site of old Fort Manuel Lisa, just above the mouth of Hunkpapa Creek, Careen County, South Dakota. This fort was built in 1812 as a depot from which to hold the loyal Missouri River Indians loyal to the American cause. It was in reality a military post built under the guise of a fur post and destroyed in 1813 by allied Indian tribes. Also Fort Manuel is the purported grave of Sacagawea (Bird Woman) guide for Lewis and Clark in 1804. There is a lot of controversy regarding whether or not Sacagawea was buried there; some historians claim she is buried in Oregon. This ferret was located on a prairie dog town which could be about 10 miles below the North Dakota line and by the way this prairie dog town also yielded many relics from an ancient Arikan village which sat on the identical site. I should state prairie dogs seem to like an area that has had a farmstead for habitat. They like ridges where old fence lines were located.

> The last one I saw was in 1958 in a dog town at a point just about almost straight south of the city dump of Mobridge, South Dakota or 2–2½ miles east of the Sitting Bull monument west of Mobridge on land at that time owned by Ted Sogge, who has since moved due to inundation by Oahe reservoir.

Ralph tallied a total of thirteen individual ferrets from 1949 to 1964, a period when he also "treated" 145,000 acres of prairie dogs with Compound 1080 and strychnine across South Dakota, North Dakota, and adjoining lands in Nebraska. After reading pieces of Ralph's life for the previous few weeks, I was crushed when I read in Ralph's last letter that he believed that "these little animals do not succumb to eating treated prairie dogs" but rather survive and just shift their diet to other things. I thought he knew his work over the past fifteen years was killing off the species he was growing to love. He had observed too much, come to know them so well, yet denied his actions had consequences like a guilty war veteran trying to justify the actions of his youth: "Walt Stammerjohn worked Rosebud and saw quite a few. Also, he worked in Carson County when I did and saw quite a few. Harvey Gibson, mentioned [he saw] six. I think he lived in He Dog village when he worked at Rosebud. I forgot to ask William Pullin how many he has seen. I'm sure Geo Baines at Custer could add on. For now they are still with us and I don't feel they are in the whooping crane, key deer or dodo bird class."

. . .

In 1964, biologists under Stuart Udall, then Secretary of the Interior, began putting together a list of rare and endangered animals, a precursor to the Endangered Species Act. That same year, Secretary Udall appointed an advisory board to investigate the federal government's role in killing wildlife, including prairie dogs, wolves, bobcats, grizzly bears, and a host of other species. Called the Leopold Board after chairman Dr. A. Starker Leopold (oldest son of the famous wildlife biologist Aldo Leopold), the group put forward what became known as the "Leopold Report" in March 1964. They acknowledged that the black-footed ferret was "near extinction, and the primary cause is almost certainly poisoning campaigns among the prairie dogs," asserting that "far more animals are being killed than would be required for effective protection of livestock, agricultural crops, wildlife resources, and human health."

By 1964, the U.S. Bureau of Sport Fisheries and Wildlife (predecessor of the U.S. Fish and Wildlife Service) was still poisoning more than a quarter-million acres of prairie dogs per year, but South Dakota still had some remnant prairie dog colonies. As eloquently described by Faith McNulty in the book *Must They Die?*, bureau employees in South Dakota like Bill Pullins were pulled in two directions. While still mandated to assist in the extirpation of prairie dogs, bureau employees also were increasingly thinking about conserving the black-footed ferret, wondering why there were so few ferrets and what could be done to increase their numbers.

Even with this newfound interest in studying ferrets, there were very few known populations left to study. The first official study of ferrets was done on Earl Adrian's ranch a few miles south of White River, South Dakota, where they had been spotted in 1964. South Dakota Game, Fish and Parks biologist Bob Henderson had a passion to study the rare mammal, but his boss allowed him to take up the hobby only if it did not interfere with his other duties. Henderson promised to keep his ferret studies under the radar, but even so, Earl Adrian didn't want a government employee on his land. Instead, he forced Henderson and his colleague, Dr. Paul Springer at the South Dakota Cooperative Fish and Wildlife Research Unit, to hire his son, Dick Adrian, to conduct the study.

Bob Henderson and Dick Adrian set up camp on a southern prairie dog colony on the family ranch and observed ferrets day and night from August 1964 to August 1965. They were the first researchers to use spotlights at night to find the reclusive animal, watch its behavior, and learn about its ecology. True to the original 1929 description by Ernest Thompson Seton, a ferret is "like a mouse in cheese, for the hapless prairie dogs are its favorite food." Ferrets never left the prairie dog colonies, only coming out of the labyrinth of underground prairie dog burrows for a few hours during the night, hiding behaviors from the researchers who could only guess how they killed prairie dogs and reared their young. On occasion, the researchers saw a female ferret move her litter across the prairie dog town single file like a "toy train." In the end, like so many wildlife studies, they raised more questions than answers, still wondering how often ferrets dispersed between colonies, and asking a question that still puzzles ecologists: How many prairie dogs does a ferret require?

At the same time, the critical question was how many other ferret populations existed. Black-footed ferret sightings became increasingly rare. Hundreds of letters came in with word-of-mouth stories of ferret-like creatures that almost always sounded more like confusion with the common long-tailed weasel. Sightings were reported in Kansas, Utah, Montana, Wyoming, and Manitoba. In the official record, the majority of ferret sightings occurred in South Dakota. Yet even in South Dakota, confirmed ferret sightings were rare. Never more than twelve in a year, and of all confirmed sightings between 1950 and 1972, more than 58 percent were from Mellette County, where Bob Henderson and the growing ferret research team focused their attention.

The following summer, Dick Adrian quit the ferret project and was replaced by Con Hillman, an ambitious North Dakota farm boy with a bachelor's degree in wildlife management from Utah State University. Con was keen to make the study of ferrets in South Dakota part of his master's degree research, and in July 1966 he found a female with a litter of five on a ranch owned by Jim Carr. Over the next sixteen months he would find sixteen more, following a total of three litters around in the night and making careful notes of their behaviors: where they moved, what they ate, and patiently waiting by their burrows to see how long they stayed underground. He reported his research as a master's thesis defended at South Dakota State University in 1968, and that same year summarized his findings for the public in the short, pioneering paper hidden on the 433rd page of the Transactions from the 33rd Annual North American Wildlife Congress unassumingly titled "Field Observation of Black-Footed Ferrets in South Dakota."

As I moved forward through the attic file boxes into 1968, 1969, and 1970, reports from South Dakota became more frequently typed and official sounding. A series of scientific publications and symposium proceedings were produced by Con Hillman, Bob Henderson, and a variety of other biologists who collaborated in Mellette County ferret research. I pulled out details from scientific papers showing that scat analyses proved that ferrets almost exclusively fed on prairie dogs. Monitoring reports showed that ferrets rarely left prairie dogs colonies and that they required sizeable prairie dog colonies of at least twenty-five acres in size, and more likely close to a hundred acres in size. All told, during the eleven years of studying ferrets in Mellette County, ninety ferrets were located, and at least thirty-eight young were produced. Yet rather than acting as a single, large, healthy population, the ninety confirmed ferrets were highly dispersed across Mellette and the surrounding eight counties—evidence that this wasn't the last stronghold but rather just the last fragments of a species on the decline.

Although the meat of the scientific reports on ferrets in Mellette County was focused on methodology and reporting results, the final concluding paragraphs began to consistently end with increased calls for ferret and prairie dog conservation. Poisoning continued throughout much of South Dakota as federal agencies contradicted each other and state and local farm bureaus continued to lobby for prairie dog control. State governors, congressmen, and even biologists continued to believe that prairie dogs were pests to be eradicated. Concessions were sometimes made to at least look for ferrets prior to poisoning—to "clear" the area and confirm that ferrets did not exist where poisoning was to occur. Unfortunately, the surveys were often done during the day and by untrained personnel with prairie dog vendettas. Even if done correctly by those who had actually seen a ferret in the wild, it would have been difficult to find a reclusive ferret during a quick prairie dog "clearance" survey. Thus by claiming to not see a ferret, biologists were able to justify their poisoning of prairie dogs on thousands of acres of ferret habitat on private and public lands. By 1968, a frustrated Robert Henderson left South Dakota. Dr. Springer was reassigned to North Dakota.

. . .

While Con and others were undertaking that first detailed study of ferrets, in October 1966, the Endangered Species Preservation Act was passed by Congress, extending full protection to thirty-six birds, six reptiles and amphibians, twenty-two fish, and fourteen mammals, including the black-footed ferret. That year, funds were appropriated for Dr. Ray Erickson, Assistant Director of Endangered Wildlife Research for the Bureau of Sport Fisheries and Wildlife, to start a captive population of black-footed ferrets at Patuxent Wildlife Research Center on the far side of the country. More than two decades before my father started work there, Patuxent was world renowned because of its history of innovative, successful captive breeding programs for critically endangered whooping cranes, bald eagles, and California condors.

Ray Erickson didn't get his first batch of wild black-footed ferrets until 1971, when six ferrets were captured from Mellette County and taken to Patuxent. Between 1968 and 1971, Patuxent staff had been practicing by raising European polecats, a stubby-bodied Eurasian ferret species that is the black-footed ferret's closest living relative. Over three years, they had remarkable success, producing more than 150 offspring and testing breeding and weaning procedures, canine distemper vaccines, and designs for nest boxes and holding pens.

Of the six ferrets transported from South Dakota, four died from a modified live-virus canine distemper vaccine before even arriving at

Patuxent. It was a vaccine that worked well on European polecats to prevent a disease now known to be 100 percent fatal to black-footed ferrets. But black-footed ferrets were sensitive to the live-virus treatment, and only males FM-71–1 and FM-71–2 survived.

On September 15, 1972, Con Hillman captured female ferret FF-72–1, injected her with a killed canine distemper virus vaccine this time, and drove her to a secure National Guard compound in Rapid City, South Dakota, for a three-week quarantine. After surviving her period of isolation, on October 4, the female ferret was transported by Con to Dulles Airport in Washington, D.C., where she underwent an additional month of quarantine at Patuxent Wildlife Research Center prior to being placed in a cage adjoining FM-71–1 and FM-71–2 on November 6.

The records I found in the Patuxent attic on ferret husbandry for the three individuals were exhaustive and repetitive. They were being fed a daily diet of 50 grams of dry mink feed, 100 grams of dried dog food, and 20 cc of corn oil—a far cry from killing a prey item that outweighs you by a quarter of your body weight. Prairie dogs bite with sharp incisors, mob you with family members if you venture out during the day, and bury you underground when you are sensed in one of their burrows. Wild lions hunting wildebeest on the Kalahari have nothing on the black-footed ferret hunting prairie dogs on the Great Plains.

But now this relationship was being lost, prairie dog without predator, predator without prey. The ferret population in Mellette County was in a mysterious and precipitous decline. Between 1973 and 1974, three more Mellette County ferrets were trapped and brought to Patuxent. After 1974, no ferrets were seen in Mellette County or in any of the adjacent counties where they had been observed in the past. Ferret searches intensified as the newly established Black-footed Ferret Recovery Team developed a recovery plan in the event that captive breeding might take off or another population might be uncovered. Hundreds of reports came in of ferrets spotted in Kansas, Montana, Wyoming, Colorado, Utah, New Mexico, Nebraska, Canada, and South Dakota. Most of these reports were of little use, unable to be confirmed or far away from prairie dog colonies where ferrets were likely confused with long-tailed weasels. Other reports were more credible, like one dated August 6, 1974, from eminent big cat expert Dr. Maurice Hornocker, who reported seeing a ferret running across Interstate 80 twenty miles west of Laramie, Wyoming, at 9:30 in the morning. All of the reports were scribbled on or stapled to a piece of paper on which it was written that they were unable to be verified or not worth following up before they were filed away only for official records. At the same time, black-footed ferrets were officially considered extirpated in Canada, as well as Texas and Oklahoma, with other states to follow soon afterward.

At Patuxent, captive breeding had stalled. The black-footed ferrets were not breeding successfully like their European polecat surrogates with which the biologists had been practicing for years. Husbandry and feeding regimes were altered and tested with almost no success. In 1976 and 1977, one of the two surviving females produced a litter each year, but all ten kits were either stillborn or died within days of birth.

At the bottom of the last box of files, the final report was dated 1979. By then all captive female ferrets had died, and only a single male remained. He perished later that year, and I thought that there could be few things as delicate as a black-footed ferret. The last individual of an entire species was housed in a building just down the road in suburban Maryland, minutes from the Baltimore–Washington Parkway, and less than an hour from the White House. The leading experts in the world on captive breeding and species conservation had failed. I looked for other file boxes with ferret records. I believed there must be more boxes, more letters, more reports. Something to continue the story, let me believe there were still hidden corners of the western U.S. where the species might persist.

I went down to my father's office. He was on the phone so I waited just outside the door, trying to listen for the end of his conversation. I was confused, distraught, I wanted to know there were still ferrets out there. I wanted him to tell me they had been rediscovered. As he hung up the phone I walked in and sat in the wooden chair opposite his oak desk.

"Has anyone seen a ferret since Patuxent?" I asked him.

"Yes," he said. "I believe there are ferret biologists working in Wyoming."

CHAPTER 3

Rediscovery

By the early 1980s, 412 preserved black-footed ferret specimens were known to exist in museums, and Elaine Anderson tried to hunt down each one. She eventually created a map of dots showing the collection points of specimens that dated back to the 1880s. The map she came up with roughly tracked the extent of the known range of three prairie dog species (the Gunnison's, white-tailed, and black-tailed), bounded by Texas and Northern Mexico to the south and by Montana and southern Saskatchewan to the north. The black-footed ferret was a uniquely North American species.

Even prior to the loss of what was known as the last wild ferret population, in Mellette County, South Dakota, in 1974, because of the large potential range map and knowledge of how secretive the species was, some people still harbored hope that another ferret population would be discovered. Tim Clark was one of them. In 1976, after completing his dissertation on prairie dog ecology, with Tom Campbell he created the Biota Research and Consulting firm—a generic title for a small and highly specialized group that focused their efforts on the goal of being the ones to rediscover black-footed ferrets.

Rediscovery of a black-footed ferret population was the kind of event that would make you a celebrity overnight. For a young scientist, it could kick-start a research program and form a funding platform that would last for years if not decades. A hunt that, because of the black-footed ferret's endangered status, had to begin with a permit from the U.S. Fish and Wildlife Service.

Mr. Ron Nowak
Office of Rare and Endangered Wildlife
U.S. Fish and Wildlife Service
Department of the Interior
Washington, D.C.

Dear Mr. Nowak:

I am currently making plans for the 1976 black-footed ferret search in Wyoming. As you know my work over the last two years was an extensive search and yielded several localities where reports of ferret sightings are concentrated. What I plan to do this year is conduct an intensive search in one or more areas. The first two areas I would like to thoroughly examine are prairie dog concentrations 1) along Powder River and its principal drainages in NE Wyoming, and 2) Big Sandy BLM area of SW Wyoming.

In a detailed report which I sent your office in late 1974 I discussed the limitations of employing the techniques used in South Dakota [spotlighting] in the sagebrush areas of Wyoming. As a result of this presentation, I would like to as part of this years search to make application to try to live trap ferrets in these two areas. I would use #202 Tomahawk live traps. These are the same design I've used over the last year on my pine marten study in Grand Teton National Park.

If a ferret were captured I would follow these procedures: 1) Wyoming Game and Fish biologists and wardens from the area, along with U.S. Fish and Wildlife personnel and BLM biologists would be notified of the find, 2) they would be invited to come to the capture site to observe the ferret, and 3) the ferret would be released in their presence as soon as possible at the capture site. This procedure would conclusively demonstrate the presence of a ferret!

Sincerely,
Tim W. Clark, Ph.D.

Over the following years, Clark and Campbell's personal mission extended beyond just Wyoming. The search for a remnant ferret population eventually led them to visit a total of eleven states, spending long nights on the prairie searching for tracks, trenches (where ferrets have excavated a prairie dog burrow by kicking dirt out a four-foot stripe of loose dirt), any kind of ferret sign. They surveyed local biologists and took to the air to search for potential large prairie dog colonies, working on the bottom-up hypothesis that if they could find a large prairie dog colony, they might just be able to find a place where ferrets were able to sustain themselves—corners of the prairie where maps were blank and others might have forgotten to check. After exhausting their hunches and with no individual success, they eventually cast their net wider by offering a $250 reward to anyone with information leading to a confirmed ferret sighting.

At the same time, federal biologists were also looking, but with the opposite intent of Clark and Campbell. Under a requirement of the Endangered Species Act, the U.S. Fish and Wildlife Service was charged with confirming the extirpation of ferrets—a political maneuver to ensure that should an area be developed, industries, landowners, and state and local governments could go about their pursuits without concern of lawsuits brought about by affecting the endangered black-footed ferret. Thus, prior to any type of land development, federal biologists like Max Schroeder, Dennie Hammer, and Steve Martin were charged with surveying and declaring the absence of ferrets. They were issued pickup trucks and dispatched to federal lands across the Great Plains to survey and confirm that nothing of legally protected status was going to be plowed over and away. They often tried to cover thousands of acres at a time ahead of the exponentially increasing demand for surface mining, the destructive beast of energy and commerce that scraped away the life of the land to get at the carbon underneath, forsaking one source of carbon for another, more densely packed and able to be burned. After what must have seemed like endless nights searching for ferrets with spotlights, effort reports were filed, sections of the map were marked off. Hope for the existence of black-footed ferrets was decreasing by one large swath of prairie at a time.

. . .

Around the same time, Michael Soulé and Brian Wilcox convened the First International Congress on Conservation Biology in 1978 in San Diego. Although the concept of conserving species had been around for decades in the United States, and for centuries in Germany and India, never before had ecologists taken such a proactive stance toward the protection of the natural world. They took a step away from hard lines of experimental science and reasoning toward a field that blurred the lines between advocacy and the more traditional scientific objectivism. As Soulé and Wilcox would later define the field, "Conservation biology is a mission-oriented discipline comprising both pure and applied science." It blended traditional biological and ecological sciences with economics, sociology, and education, all in an effort to preserve species in the face of human persecution that was happening at local and global scales.

The urgency of the field was unique, controversially setting a framework within which scientists could enter into the realm of management for the sake of protecting what they study. As tropical ecologist Daniel Janzen would later advocate, "If biologists want a tropics in which to biologize, they are going to have to buy it with care, energy, effort, strategy, tactics, time and cash. . . . If our generation does not do it, it won't be there for the next. Feel uneasy? You had better. There are no bad guys in the next village. They is us." Scientists could no longer take the high road and claim that extinction is a natural process, part of Darwinian evolution, and move on to studying the next biological phenomenon.

For a discipline that emphasizes the conservation of biological diversity (the term *biodiversity* would not be adopted until 1988), *extinction* was the dirtiest of words. Many of those attending the San Diego meeting had worked on island systems or were still focusing their thoughts on the theory of island biogeography that had been developed by Robert MacArthur and E. O. Wilson a decade earlier. The influential theory was that the risk of extinction for a species on an island is a function of the size of the island and its isolation from other suitable islands. There was a corresponding belief that on the mainland, habitat fragmentation (in which humans removed habitat and further divided the landscape) similarly affected species. If we isolate a species on one of these virtual islands, such as a patch of forest surrounded by corn fields, the probability of that species going extinct increases. The smaller the patch, the worse off a species would be.

To avoid losing these "island" populations and risking extinction, ecologists began trying to identify critical population sizes to avoid extirpation. Mathematical models were developed for an individual species or population that, in their simplest form, were projections of the population into the future based on historical rates of births and deaths. But there was a need to account for a number of obstacles that could make small populations prone to extinction. Small populations within an island of habitat are more vulnerable to unpredictable events like floods or catastrophic fires.

For his dissertation research in 1978, Mark Shaffer developed a formula for predicting the minimum population size of grizzly bears to withstand such unpredictable events, which he termed the "minimum viable population size" (MVP). Over time, others would add to the complexity of such MVP models by including long-term detrimental effects like decreased reproductive output as a result of inbreeding and other factors. The product was hard numbers that could be used by wildlife managers to outline how many animals they needed to maintain, similar to agricultural balance sheets used by cattle ranchers to balance predicted losses against predicted gains, trying to stay in the black. But more than that, by experimenting with the numbers in the model, scientists could begin to say what factors needed to improve to keep the species viable. How would the population respond to a year of good rain that would perhaps result in a sudden 30 percent increase in juvenile survival? What would happen if we stopped a hunting season and were able to decrease adult mortality by 10 percent? These were hard numbers rather than sentiments to bring to politicians and the public when trying to maintain or recover a species.

The risk of species loss and the value in conserving rare populations also were brought to the forefront of public policy. The Endangered Species Act of 1973 (built on the backs of previous acts of 1966 and 1969) was amended to require not only listing of threatened and endangered species, but also designation of critical habitats and development of population recovery plans for those listed species. The National Forest Management Act of 1976 required the U.S. Forest Service to "provide for diversity of plant and animal communities based on the suitability and capability of the specific land area." In effect, this made national forest managers responsible for maintaining viable populations of native species in their planning areas. Given that endangered and threatened species were typically most at risk of loss, the emphasis was put largely on maintaining or increasing their small populations.

The theoretical and political pieces were coming into place for the conservation of black-footed ferrets, should they ever be rediscovered and brought back from extinction.

. . .

On the morning of September 26, 1981, John and Lucille Hogg were having breakfast in their ranch home, an unassuming white-painted house in the center of a moderately sized ranch in the open, rolling hills and sagebrush flats west of Meeteetse, Wyoming. Up and down the valley the bustle of summer had ended. The rush of growing, irrigating, and cutting hay for the winter had given way to winter preparation. The cows and their calves were coming back into the corrals on homesteads and off the federally leased summer range. Daily chores switched from fences and farming to feeding livestock and preparing for the calving season, assessing the summer's alfalfa hay crop, and calculating whether there was a need to buy hay to make it through the winter.

The ears of a sleeping rancher must be finely tuned to the sounds of the outdoors, keen to the sound of a bawling cow in early labor, even through the walls of a house in the middle of the night. Lucille Hogg hadn't slept well because their Australian cattle dog, Shep, was barking and growling outside in the middle of the night. Lucille tossed in bed, settling on the thought that the dog was just learning another lesson on why to steer clear of porcupines. Knowing that there was little she could do in the middle of the night to pull out the hooklike barbs, she decided it could be dealt with in the morning, so she fell back asleep.

Over breakfast, Lucille told John to check Shep for porcupine quills and inspect the damage before the quills corkscrewed too deep into his muzzle and a trip to the vet was in order. Stepping out the front door, John noticed Shep was intact and uninjured, ready for the day. On the front stoop he'd left a carcass, the likely victim of his nighttime ruckus left for show to his owners. As John peered down at the thin carcass, his first reaction was that he had no clue what Shep had dispatched. It was a buff-brown animal with black markings on its paws, face, and tail. Perhaps it was an oddly marked pine marten that had come down from pine forests of the Carter Mountains to the north. Regardless, it looked interesting and attractive enough to pay to have preserved, and John took the specimen to Larry LaFranchi, the local taxidermist. Having stuffed more than his fair share of pine martens, Larry immediately recognized the animal as a black-footed ferret and called Wyoming authorities.

A ranch dog had succeeded where so many professional and backyard biologists had failed, offering a humbling and at the same time hopeful sign that the human footprint had not touched everything. Shep's singular finding hinted that there were places left forgotten, little understood, where even the most sensitive of species could still persist. A second chance to perhaps learn from these animals and make up for our errors of the past, maybe even resurrect the species.

Within days the State of Wyoming held a town meeting in downtown Meeteetse, where rumor of the find had spread. Why had black-footed ferrets persisted here, on the very western boundary of the prairie? A single isolated dot on the western edge of the historic range map Elaine Anderson would develop that was based on all known museum specimens. Species are supposed to decline from the outside in, shrinking down into the middle where a small refuge remains. This small town of

Meeteetse, Wyoming, just south of the more famous Cody, Wyoming, and Yellowstone National Park, was outside the area most thought to survey. But questions of why were secondary to questions of whether there were more, and how many. Wyoming Department of Game and Fish biologists started the meeting by reporting the details of the find and asked the audience if anyone had seen a ferret. Immediately Doug Brown volunteered that he had seen a ferret while working on the Pitchfork Ranch.

In the small ranching community of Meeteetse, stories of where ferrets remained were mixed with fear of what having an Endangered Species on your property could mean. There were rumors that federal restrictions could follow, brought on by enforcement of the Endangered Species Act—the type of legal maneuvering conservation groups still try to use to force the federal government's hand to pressure private landowners about the use of their land. Despite that the precedent was rare and that most induced management changes occurred on federal land, the fear tactics create a persistent tension between conservation groups and private landowners, with the federal government in the middle. For independent ranchers, relying on the federal government as an effective moderator often wasn't enough. Policies and administrations changed; ranchers were forced to live with the consequences. Despite the circulating paranoia, with counterculture bravery and vision (traits that, in future years, would be evident in almost every ferret conservation milestone), Jack Turnell, the owner of the Pitchfork Ranch, the Hogg family, and a handful of other ranchers in the vicinity ignored public concerns and allowed researchers onto their land.

Tim Clark and Tom Campbell rented an airplane to map the extent of prairie dogs in the area. They scanned the rolling, open sagebrush steppe grasslands of the Big Horn Basin around Meeteetse that are bounded to the west by the Absaroka mountain range that extend north into Yellowstone National Park. On the ground, federal biologists Hammer and Martin were conducting nightly searches for ferrets on the group of ranches adjoining the Hogg ranch. They walked or drove around the prairie, using spotlights to look for the distinctive green eye shine from the reflective membrane just behind the ferrets' retinas, hoping that a curious ferret would stick its head out of a burrow. To improve the chances, they also used two search dogs trained to sit when they smelled the scent of a ferret. When one of the dogs smelled the faint aroma of ferret scat or musk and sat down by a burrow, Hammer and Martin, not seeing ferrets themselves but blindly hoping the dogs were on to something, would set live traps at that burrow and surrounding burrows.

On October 29, 1981, more than a month after Shep's discovery, Hammer and Martin were checking traps in an area where dogs had found scent. Leaving the traps set, they returned to their vehicle and drove off into the night to continue their spotlight search. Looping back around toward their traps, at 6:20 A.M. they saw the flash of something running across the road in front of them and periscoping its head out of a burrow. They had never seen a ferret, but they thought it looked about right; with buff markings, it skimmed quickly across the prairie like a weasel relative. Hammer approached the burrow, and the ferret was still just inside the entrance, peering up from the shadows and chattering at him.

No human could take credit for rediscovering the species—that honor belonged to Shep—but few could have experienced the emotions of Dennie Hammer and Steve Martin. They experienced the career-defining moment of being the first people to see a live black-footed ferret in two years, after the last male died in captivity at Patuxent and the species was widely believed extinct. Perhaps more important, they were the first to hear the chatter of a live ferret in the wild since Mellette County seven years prior. Adrenalin mixed with panic as Dennie put his hat over the burrow while he rushed to get and set a trap. Walking up to the trap eleven hours later and seeing the ferret looking back at him, his thoughts must have been filled with that delicately marked prairie bandit. It was the almost mythical creature for which he had searched hundreds of hours across much of the western United States. This erased the ridicule from other university students who, when he said he wanted to study ferrets for his graduate degree, told Hammer he was trying to study dinosaurs. Then there was the thrill the next day of releasing the ferret that they had nicknamed "620" (after the time it was captured). They watched it scurry down the burrow after it was fitted with a small radio-transmitter collar and hoped it would stay alive so they could monitor its movements. There was so much to learn. Did ferrets here in the shortgrass prairie of Wyoming behave differently than those in Mellette County, South Dakota? Why were they here and nowhere else? And most important, how many were there? Now, finally, ferret research and conservation could begin again in earnest.

. . .

Hammer and Martin, along with telemetry specialist Dean Biggins and a handful of additional federal researchers, intensively monitored the movements of ferret 620 for the next month. They eventually found ten other ferrets in the area. Some of the mysteries of the black-footed ferret had been unlocked in the final days of research in Mellette County, South Dakota. Con Hillman and others had noted that ferrets were typically nocturnal and solitary, mostly visible in early morning or late evening. They found that adult females, on average, produced 3.4 young each year, and that the young typically first venture above ground as early as mid-July. They also learned that ferrets were most readily spotted in early fall when kits dispersed and mothers and their kits played above ground. They learned that ferrets were underground much of the winter, limiting biologists to monitor infrequent movements based on ferret tracks and diggings that could be spotted after a fresh layer of snow had fallen.

But there was still so much to learn. Because only a few ferrets were followed in Mellette County and none was observed to move between prairie dog colonies, it still was not known how and when ferrets moved between colonies. Little was known of their breeding behavior, whether females reared young alone or had help from males, how often they fed their young, or how many prairie dogs were required to sustain them. Were ferrets such efficient predators that they served as a sort of natural control for prairie dog populations, as C. Hart Merriam had predicted in 1902? Given that ferrets vanished from Mellette County, was a particular size and arrangement of prairie dog colonies required to sustain ferrets? There were still the questions of what role prairie dog poisoning had had on ferrets, and how many prairie dogs were required to support a self-sustaining ferret population.

Further, it was possible that ferrets in Meeteetse might behave differently than those in Mellette County because this newly discovered population was on the edge of the ferret's supposed historical range. The ferrets in Meeteetse persisted on a colony of white-tailed prairie dogs, unlike the ferrets in Mellette County that lived on black-tailed prairie dog colonies. White-tailed prairie dogs were arid-environment specialists, primarily appearing in portions of Montana, Wyoming, Utah, and western Colorado. Lower rainfall meant less vegetation in both summer and winter, resulting in a species that evolved to live in more widely distributed social groups and lower densities. How did ferrets persist on sites having less prey? Did they need to travel farther, were they more competitive, and did they supplement their diet with other rodents or birds?

It was the advent of radio-tracking technology that allowed researchers to begin addressing many of these detailed questions for such a small, cryptic species. By the late 1970s, technology had advanced to the point that small transmitters could be attached with cloth collars to animals. Biologists had already identified the size, design, and weight of a radio collar that the close relative of the black-footed ferret, the Siberian polecat, could tolerate. They knew that ferrets would present a unique issue compared to other animals. Their small weight limited how large a transmitter they could tolerate without the ferret's ability to move and hunt being hurt. Biologists knew that without a large battery, the transmitter would have limited range and lifetime. They knew that because a ferret spends more than 90 percent of its life below ground, getting a signal would be difficult at best, and it was more likely that they would have to scan continually, often in the middle of the night, for the chance to detect that a ferret had come above ground.

In addition to the technology encased within the plastic-coated transmitter capsule, for a tubular animal the design and tightening of the collar itself was critical. A collar needed to be loose enough to allow breathing and growth of the animal, tight and rugged enough to stay on, and made of cotton that would degrade so the collar would fall off over time as the battery died. This made the stress of placing a collar on a captured and sedated ferret that much more intense. Not only did you have one of the rarest animals in the world in your hands, you had to mount something around its neck that if done improperly could kill it. If it was too loose, you would risk the chance of the ferret slipping the collar, forcing a second stressful capture if you ever even saw the animal again.

For wildlife biologists, tracking an animal is typically a solitary affair conducted by a biologist with a hand-held antenna and receiver tucked under the arm, headphones sending out a hissing sound. The biologist listens for a faint beep as the antenna is panned around, and hikes from waypoint to waypoint, eventually honing in on the location of the animal and taking note of the location. Yet, because ferrets came above ground so infrequently, just detecting them required teamwork. Rather than sending a dozen biologists out onto the prairie throughout the night, three camper trailers, modified so that large directional antennas poked through their roofs, were placed on the highest points surrounding core prairie dog colonies in a triangular design. This required near-constant tracking by a team of biologists who spent hours hunched over within the trailers, rotating the antennas and scanning ferret transmitter frequencies, listening through headsets for faint electronic beeps in the hiss of static signaling a ferret had come above ground. Once a ferret emerged and was detected by one of the biologists, the biologist would relay the approximate location of the ferret to biologists in the other two trailers. When two or more trailers locked on to a ferret signal, they could accurately triangulate the location of the ferret. Over the span of weeks and months, researchers began to define the territories of individual animals, their activity patterns, and their behavioral interactions with other marked ferrets.

As a break from the trailers, researchers would swap their sedentary nightly duties and hit the sagebrush flats to find ferrets by spotlighting, searching for isolated animals on the peripheral colonies that were not already under radio-telemetry study. The researchers would move around on the prairie and capture the remaining few ferrets, recording their measurements and marking them with "passive integrated transponder" implants (PIT tags for short) injected just below the skin. These small, pill-shaped microchips are passive in the sense that they do not rely on a battery to transmit a signal like radio collars do, but instead require an outside scanner to pass within a few inches to relay a reading. With no bulky batteries to worry about, PIT tags last forever and provide a way to permanently identify individual ferrets. By finding and scanning the PIT tags of ferrets over the entire area, biologists were able to get a count of marked individuals. They could learn a minimum population size with which to assess the status of the population and follow the population through time to assess its viability.

With these tools, the small army of researchers learned that it was unlikely to see two or more ferrets together. The solitary ferrets seen above ground in January and February were likely males in search of mates, roaming their territory that overlapped one or more smaller territories of females. Seeing one male trying to maintain access and breeding rights to multiple females suggested to researchers that ferrets were similar to other members of the weasel family in having a polygamous mating strategy. Ecological theory tells us that in polygamous societies, a male has to produce as many offspring as possible to maximize his genetic lineage. This requires that males keep exclusive access to as many females as possible. It is a matter of quantity, but also quality, because females typically select the best habitat in which to raise their litters. Thus, by association, males that protect prime habitat also are likely to have access to the most females and to females that are most likely to successfully rear their young.

Guarding of mates by male ferrets reached a fever pitch in the mating season from January through mid-March. Then, after a forty-five-day gestation period, kits were born in a natal burrow. Contrary to many other carnivores, such as wolverines and otters that remain in a single nest chamber for the whole litter-rearing period, female ferrets would move kits to new burrows at regular intervals. The mother would carry them with her teeth by the scruff of the neck, one at a time, when their eyes were still shut and they were not yet mobile.

Meeteetse researchers found that females did all the caring for the young, protecting them from birth in spring through to independence in late summer. A mother ferret at first nursed her kits, staying below ground for long periods of time. She killed prairie dogs only occasionally for her own food, until June, when she started moving the whole family to a burrow containing a freshly killed prairie dog. By late July and early August, kits were old enough to have motor control and come above ground for periods of time during the night. They would be cautious at first, with only the boldest of a litter of three or more doing more than sticking its head above ground, and never straying more than a few feet from the safety of the natal burrow entrance.

By mid-August, kits came above ground almost every night. On 93 percent of nights to be exact, typically between 1:00 and 4:00 A.M. The kits stood by their natal burrows as the mother hunted for food, until the silence was broken by one of the litter starting a tussle by charging at another, back arched, mouth agape, tail frizzed. Spotlighters found the kits to be curious, peering up from burrow entrances when the spotlighters approached. The kits noticed that any foreign object placed by their burrow deserved inspection. The small flags the researchers used to mark burrows often induced a spontaneous fit of play. Young kits would lunge off the ground, reaching the flapping flag at the top of the two-foot wire flags and landing in a puff of dust. They seemed amazed that they reached so high after a childhood spent mostly below ground or within two inches of the flat prairie surface. Collecting themselves after landing, as the dust cleared they sometimes chased their own tails, perhaps out of curiosity about their rapidly growing bodies.

By late August, kits grew to be as large as their mother or even larger, and more independent. They dispersed from their mother's territory by the end of September and the prairie dancers had gone away. They were solitary for the next few months as behaviors shifted from rambunctious play to the pressures of adulthood and survival. By November, the former kits and older adult ferrets were active above ground only for, at most, one to two hours per night. With so little to observe, researchers switched their monitoring from spotlighting to occasional snow tracking when fresh layers of snow fell, searching for slight pawprints leading between burrows in skiffs of thin snow blown like sand into small waves by the bitter prairie winds. Biologists found that during the harsh winter months, when temperatures dropped below freezing and when white-tailed prairie dogs slept below ground for weeks on end in a state of torpor, ferrets similarly slowed their above ground activity. During peak winter periods, ferrets would spend up to six nights and days below ground without moving to another burrow. But the clearest pattern to the researchers was also the one attribute that has been known the longest about this species since its first discovery and description: that black-footed ferrets almost never leave prairie dog colonies; as the prairie dog goes, so goes the ferret.

. . .

With the birth of conservation biology as a major scientific discipline, its participants formed a society in 1985 and soon thereafter launched an academic journal. The very first issue of the journal, *Conservation Biology,* published in 1987, opened with a four-page "progress report" by Tim Clark on the conservation of the black-footed ferret. He stated authoritatively on the first line that "black-footed ferrets are the most endangered mammal in North America."

The remainder of the article painted a similarly grim picture. For what Soulé termed a "crisis discipline," the black-footed ferret story of Meeteetse served as an ideal case study of conservation biology in practice. The series of events following rediscovery provided the classic example given in textbooks for the next several decades about how conservation biology requires skill sets from many walks of life to restore critically endangered species from the brink of extinction. And examples of ferret recovery efforts also demonstrated how failure to gain broad support and consensus on overall management direction can result in near-catastrophe.

In 1982, the year after rediscovery of ferrets at Meeteetse, researchers counted sixty-one individuals. That number increased in 1983 to eighty-eight and then again to 129 in 1984. The population was thought to be productive enough to exceed requirements of a minimum viable population. This meant that some ferrets could be captured and used as seed stock for captive breeding at the National Zoo, and once again at Patuxent. Yet because of political infighting between the State of Wyoming and a host of researchers and federal agencies, Wyoming decided that ferrets should not leave the state. Because no adequate facility existed in Wyoming to keep ferrets, let alone breed them in captivity, and because Wyoming insisted that any new breeding facility should be paid for by federal and private sources, capturing ferrets for captive breeding was put on hold.

Unfortunately, surveys by Meeteetse researchers found that by August 1985 there were only fifty-eight ferrets. Biologists feared the worst as the population continued to dwindle to thirty-one by September and to sixteen by October. There was great confusion over the cause of such a precipitous decline. Earlier that year plague was reported in the area, but studies of European polecats suggested that ferrets were likely immune to plague (an assumption that later turned out to be wrong; black-footed ferrets are actually highly susceptible to plague, as discussed in later chapters). Regardless, such a precipitous decline made the ongoing debates among state and federal biologists moot. The race was on to save the last few individuals in a last-ditch effort at captive breeding.

In October 1985, six of Meeteetse's remaining ferrets were captured and transferred to a Wyoming wildlife research facility located in Sybille Canyon in the southern part of the state. Upon arrival, one ferret died of canine distemper virus. Then another died. Finally, because all six were housed in the same room at Sybille, the remaining four eventually contracted distemper and died.

A capture team was immediately sent to collect all remaining ferrets from the Pitchfork Ranch. Six more ferrets were brought to Sybille the following week. These animals did not die of distemper, but six individuals was hardly enough to start a captive breeding program. Certainly, such a number gave the captive breeders and ferrets only a small margin for error. But researchers knew that there was only a small chance that any more ferrets could have evaded capture and survived the canine distemper outbreak that was known to have spread through Meeteetse that year.

However, surveys in 1986 revealed that four individual ferrets survived in the core of the Pitchfork Ranch. There were two males and two females, which were monitored through the summer and found to produce litters of five kits each. Despite this glimmer of hope, by March of the following year it was decided that all known ferrets should be captured and moved to Sybille. In all, eighteen surviving ferrets were captured and taken into captivity. Some stragglers might have avoided this final capture effort, but they likely succumbed to disease or natural mortality in the coming winter or spring. A few might have lived as long as a year or two, but they would have been too few in number to persist and find each other to reproduce. All we know is that no more ferrets were seen in Meeteetse after March 1987.

With the last wild black-footed ferret at Meeteetse captured, something else was lost. Tim Clark paid the Hogg family the $250 reward he had promised for a confirmed ferret sighting, and there was still hope that rare ferret family groups remained hidden in remote pockets of the American West. During the thirty years since the Meeteetse rediscovery, the reward was increased to $5,000 and then to $10,000, but no remnant wild ferret populations were discovered. We now know that Shep found the last black-footed ferret population, and the species would likely have truly gone extinct, unnoticed without his help. Decades of searching has told us with near certainty what biologists already feared as they removed the last individual from Meeteetse: there were no other black-footed ferrets on the Great Plains. The fate of the species rested in a small number of captive individuals.

CHAPTER 4

Captive Breeding

US Route 191 crisscrosses the Rockies in a nearly straight line, heading north from the Mexico border at Douglas, Arizona. It is a 1,905-mile span of asphalt that passes through Arizona, Utah, and Wyoming before stopping at the south entrance of Yellowstone National Park, then starting up again at the west entrance of Yellowstone in Montana and finishing at the Montana–Canadian border. The last town of consequence before the border is Malta, located where north–south Route 191 intersects east–west US Route 2 just below the 48th parallel. An intersection of consequence because the similarly impressive Route 2 traces the Canadian border for 2,192 miles from Houlton, Maine, on the Atlantic coast to Everett, Washington, on the Pacific coast. Along its path, where Route 2 passes through the open plains of eastern Montana, it is called the Hi-Line, a flat, sparsely populated 402-mile east–west stretch of the globe named by the Great Northern Railway, which started bringing in immigrant families and exporting grain and livestock in 1887.

The Homestead Act of 1862 provided settlers with title to 160 acres of Montana land provided they built a house, planted a crop, and maintained five years of residence. Brushing aside Native American claims to large swaths of the Great Plains, this federal law pulled in homesteaders from the eastern United States as well as Europe. Towns with names like Glasgow, Havre, and Malta were named by the Great Northern Railway to entice residents from Scotland, Sweden, Norway, and other countries farther east to the promise of fertile land and pleasant climates. Other names were likely more utilitarian or locally accurate: Cut Bank, Shelby, Wolf Point, Poplar, Bainville.

Formerly just known as Rail Siding Number 54, Malta was named by a Great Northern official in 1890 when a post office was established. The land around Malta and south to the Missouri River (currently known as Phillips County) was settled later than other parts of the Hi-Line, even those farther west toward the mountains, because it was not suitable for farming wheat as had been promised by the government and railroad pamphlets. In addition to the poor soils, short growing seasons, and brutal winters, families soon found that 160 acres was not enough even for raising livestock. In response, Congress began increasing homestead allotments to 320 and then 640 acres. By 1916 came inevitable drought that pushed many homesteaders to leave or sell off their titles to other families.

Driving any direction out from Malta you see the bones of old homesteads: dirt-floored, crumpled-over log cabins on otherwise open patches of prairie. They are signs of a more active and populated past on the Hi-Line compared to the current sparse matrix of much larger ranches or federal lands. For the fortunate or determined neighbors who were able to consolidate their land holdings, the Taylor Grazing Act of 1934 boosted cattle ranching by opening federal lands to grazing by privately owned cattle. The land had reached a somewhat stable equilibrium with man—forcing livestock raising instead of wheat farming on the people. Families had to rethink the hundred-acre farms from their homelands and live at thousand-acre scales. For survival, acceptance of this isolation by distance had to be resolute and passed down bloodlines, because individuals might come and go, but families, property lines, and federal grazing leases had to stay intact. This way of life has persisted with few changes over the past eighty years.

. . .

The first time you meet Randy Matchett is likely to be on a dirt road or prairie field camp in remote central Montana just south of the Hi-Line. He is easy to spot in his white oversized pickup truck packed full of tools like a Swiss Army knife. He always carries with him chains and fence posts to tug himself or others out of the mud, a full mechanic's chest of tools to fix any type of engine (from airplane to generator) on the fly, enough medical gear to serve as a veterinary lab, a rifle to sample coyotes for canine distemper virus, and a sleeping bag and pillow in the

FIGURE 3. Short-horned lizard *(Phrynosoma hernandesi)* on the mixed-grass prairie of central Montana.

FIGURE 4. Old homestead in southern Phillips County, Montana.

back seat to serve as his mobile home. These are tools of the trade for the lead wildlife biologist responsible for the million-acre Charles M. Russell National Wildlife Refuge. When he steps out of his truck, the first things you would notice are his felt cowboy hat and neatly trimmed, horseshoe-shaped dark mustache around a wry smile. He wears a government-issue uniform of brown denim pants and a button-up beige shirt with the U.S. Fish and Wildlife Service insignia of a flying goose on his shoulder. As he moves his wiry frame you will notice a slight arch in his back and limp that he attributes to a high school "rodeo" accident.

I first met Randy in 1998 at the University of Montana when he gave a guest lecture on black-footed ferrets. He talked to a group of wildlife biology students about how he had spent the past four years trying to restore some of the Sybille captive stock into central Montana. I was enthralled by his story, but like most University of Montana students, I dreamed of studying the larger mammals—wolves, grizzly bears, or moose—in the mountains. I jotted down his email on the corner of a page in my spiral notebook, and did not think about him again until a few months later when I was graduating and without a job. Lost in the pile of applicants who wanted to work in the mountains, Randy put me in touch with a graduate student who paid me to do a summer job of trapping prairie dogs in central Montana. We translocated hundreds of prairie dogs onto Randy's refuge to restore populations, and when that three-month job ended and Randy saw that I was still interested in the prairies, he offered to keep me on—a sort of on-the-job screening process that led him to offer me free housing and wages of $16 a day to help monitor his small population of captive-reared black-footed ferrets. I left my small car at the end of the pavement where the government truck Randy had left for me waited, and drove the sixty miles on dirt roads to a field camp he termed "Ferret Camp" that consisted of a handful of small camper trailers parked on an isolated tip of land along the Missouri River.

I was hired to follow the nightly movements of the small group of ferrets Randy was carefully trying to rear in the wild. He trusted his precious animals to me and a fellow drifter, Pete, an out-of-work high school teacher from Kansas. The goal was to keep track of how many of the handful of new wild-born litters would survive, and try to predict how many ferrets Randy might expect to see the following spring. The survival of a large number of kits meant that the Montana habitat was good. Also, the more ferrets that survived, the better Randy's chances in getting more ferrets for release from Sybille stock the following spring.

Like all field biologists, we shifted our lives to that of our study animals: sleeping days and working nights; eating one large meal a day; learning the feeling of a drop in barometric pressure ahead of a storm; judging the time of night by the height and phase of the moon.

We followed the kits every night into the fall, naming each one and tracking its movements between prairie dog burrows. We were collecting points on a map to create a sketch of each ferret's life history for Randy. We rationed our food to be able to persist on a trip to town for food every two weeks and kept up our nonstop routine until the first week in November, when freezing water lines and winter started forcing us out of the camper trailers and back to the hardtop life of highways and electricity.

Pete left camp first, ahead of the north winds of the first severe cold front. Days shortened and nights grew so cold that I had to bring my truck battery into the trailer to be sure to have enough spark to start the old Dodge the next night. Like a failed homesteader, I felt the forces of weather and solitude pressuring me to leave Montana, yet I had no job to move on to. I dreaded the sharp cut of work that is field biology, from a life of continuous motion to sudden inaction when the fieldwork ends, money runs out, or the animals move on—a separation that was all the more paralyzing because we were working on such a rare and fragile animal, not knowing whether they would make it through the next five months of winter. Whether they would be here when I returned.

By mid-November a call came through on the old camp phone that was stacked on a piece of plywood in the corner of the barn. Grabbing my hat and gloves, I rushed outside and felt ice crystals instantly form in my nostrils. The chill of the cold plastic phone on my bare ear made me recoil, forcing me to hold the phone at a safe distance. On the other end of the line, the voice of a woman from the federal government repeated my name. She seemed pleased to have tracked me down in response to the application I filed the previous year to join Peace Corps. She offered a two-year contract as a biologist in the Philippines, a free plane ticket to a country I barely knew existed. I traced my mental globe to somewhere in Southeast Asia, the Pacific, somewhere warm—I did not hesitate.

. . .

In 2002, following my two years of international service and building a life as a tropical biologist, I found myself back in central Montana. Randy offered me a full-time job to help him again with black-footed ferrets, and I accepted before even getting off the plane back into the United States. He wanted me to move up to a little waterfowl refuge on the Hi-Line of northern Montana and work full-time to help him breed and rear black-footed ferrets in Montana so that they could be released there. It was a permanent job with pay, something scripted and directed, yet we knew that the outcome for the species was still far from certain—extinction loomed.

I arrived in the small town of Malta during spring, a time of hope and activity—hope that the ephemeral prairie rains would come and last into July before the grasslands dried out and returned to their more familiar and less nutritious brown. It was a time of preparation ahead of the busy summer months when ranchers must push cows out into summer pastures, fatten calves, and harvest hay before the long winter. But in Malta I was not a homeowner, not a landowner, not even a renter, more of a squatter. I was a twenty-five-year-old wildlife biologist living in a retired FEMA trailer left over from the latest Gulf of Mexico hurricane disaster that was government surplused and then hidden behind a maintenance building on the seldom-visited Bowdoin National Wildlife Refuge—a pinprick of a refuge on the larger map of federally protected areas barely large enough to justify having a full-time staff. I often imagined the poor family who had lived in the trailer before me. I created an image of a family from Alabama or Mississippi who had lost everything, cramming four kids and a grandmother into one of many white, tightly packed, federally purchased trailers in an abandoned soccer field on the Gulf Coast, each filled with a traumatized family under the government's care. Based on the stains on the sofa and smell of the curtains, I thought that my trailer's family must have had to stay in the encampment for months before they could find a modular house to rent with federal disaster aid checks, likely settling in a town far from the memories of the coast and where they had few relatives.

For me, the trailer was temporary living turned semi-permanent by cinder blocks under axles and power from an extension cord. The eyesore was hidden from public view by a ring of invasive, pale, stunted, and thorny Russian olive trees—few native tree species could tolerate the winter cold, summer heat, poor soil, and infrequent rain. As I cooked dinner on the cramped two-burner camper stove with borrowed pots and pans, I was given time to think, to reflect on how quickly my life became focused on a similar trailer three hundred feet away where dozens of beady eyes of baby black-footed ferret kits were still hidden from light by sealed eyelids. My list of chores for the next day accumulated in the back of my mind: clean nest boxes, weigh kits, disinfect floors, order implant tags, clean hamster colony, order new water bottles, mow pre-conditioning pens, feed ferrets, defrost tomorrow's food.

In the evenings, memories of saltwater ocean and love came forward only when I finished my third bottle of beer. My thoughts and emotions of loss were likely not so different from the homesteaders from Europe who arrived in Malta more than one hundred years prior to me. Like them, the promise of the open plains drew me. I had headed west at the first chance for teenage independence to work summers on a Wyoming cattle ranch with my uncle. Spent four years at the University of Montana being trained as a wildlife biologist, and then traveled around the world once again to end up in the plains of central Montana.

Yet upon returning to Montana, the critical satisfaction I had found in its open spaces was gone. I had first met J after four months of living with hunters in the Philippine jungle. While I was digging with a pick-axe to build a hiking trail, she arrived at the wildlife sanctuary, blonde and confident on the back of a British United Nations volunteer's motorcycle. She left with the British volunteer but came back a week later, tracked me down, enticed me to leave my post on occasion and travel the island, and made me fall in love. We had a tropical island love as only two biologists can, measuring the jungle trees, mapping the extent of coral reefs, surveying fish in the weekend market; reporting on components of natural beauty and living in thatch huts, hers on the coast, mine in the interior jungle. A life of continual motion in a corner of the world, without family so that we grew reliant on each other, yet knew there was an expiration date on our temporary lives when the two-year contract ended, a make-or-break deadline we avoided until the end, when she was first to leave. Yet nothing ended, and when forced to make a decision on the future, we made promises of finding each other back in the United States like high school sweethearts going away to college.

After dinner I went outside to smoke my next-to-last Philippine cigarette, knowing I needed to not be an addict and hoping to leave nicotine and the longings of my past in one symbolic, sweeping gesture of my body and mind. Outside the screen door of my Malta camper, I couldn't focus on J or the tropics. I couldn't confuse myself with the persistent question of why I chose to move to Montana, Randy, and black-footed ferrets rather than follow her to the East Coast. The mosquitoes were kept from my nose and mouth by the smoke, but they swarmed my ears, hair, arms, and legs. The mosquitoes of the Hi-Line were worse than those of any tropical forest or swamp. They swarmed on this temperate plain with the spring pulse of water to the point where you inhaled two or three with each breath. They turned the backs of white shirts grey, and then spotted with red where you had slapped one midmeal. I learned to walk fast, from door to door, car to building, home to office, to avoid the growing hemoglobin-seeking cloud from catching up to me as they honed in on my carbon signature. I finished only half the cigarette and retreated into my trailer, spending the next five minutes killing any individuals that followed me inside.

• • •

After the crash of the last wild ferret population in Meeteetse in March 1987, a captive population totaling eighteen ferrets was established at the Wyoming Game and Fish Department's Sybille Canyon Wildlife Research facility. This included the critical late addition on March 1, 1987, of Scarface from Meeteetse, a particularly virile male who helped breed nine of the eleven females. Of the nine bred females, only two produced litters in the spring of 1987, eventually resulting in seven surviving young. It was a surprising success compared to the experiences at Patuxent a decade earlier.

In 1988, thirteen of the fourteen females produced litters, resulting in thirty-four young. There was high hope for the captive breeding program because minimum production numbers seemed to be met. It was at least a small buffer from extinction. Within a few years, the number of ferrets in captivity increased enough to allow the captive population to be subdivided. Individuals were shipped to the National Zoological Park in Front Royal, Virginia, and Henry Doorly Zoo in Omaha, Nebraska, dividing the population and limiting the risk that a disease outbreak or other catastrophic event at one site would cause extinction of the species. Eventually, ferrets were also housed and bred at zoos in Phoenix, Toronto, Louisville, and Colorado Springs.

Captive breeding was so successful that by 1991, ferrets were beginning to be released into the wild. Demand by states quickly exceeded supply, so a movement began to start small-scale captive breeding programs at the state level for those states that wanted ferrets. Captive breeding buildings and preconditioning pens were built in Colorado, Montana, New Mexico, and Arizona. By 2002, when Randy brought me back to the Hi-Line, the newly created Montana ferret facility had a couple of years of experience under its belt. Randy had already hand-picked as his captive breeding team leader Valerie Kopsco, a thin, energetic biologist from New Jersey with a soft spot for cowboys. The previous year, she had already tested the specially designed outdoor pens and indoor cages with a few trial ferrets. I was brought on as a second hand ahead of the big push to finally breed ferrets successfully and raise young kits in Montana.

We followed the husbandry protocol developed by the ferret recovery program on the basis of its success with the last eighteen individuals from Meeteetse. A sort of how-to guide for ferret keeping, this protocol had evolved over time with guidance from the Wyoming Game and Fish Department, the Captive Breeding Specialist Group of the IUCN (International Union for Conservation of Nature), and the American Zoo and Aquarium Association, and with hands-on expertise of many devoted biologists over the years at Sybille Canyon, Wyoming. By 1996, the U.S. Fish and Wildlife Service had assumed responsibility for captive breeding from the Wyoming Game and Fish Department, and the person to contact about any captive breeding question was Paul Marinari. Paul would be the first to deny the label of "Mr. Ferret," placing credit with the teams and individuals that preceded him, but he more than any other person oversaw the quick acceleration in ferret-breeding success, and developed a smooth and effective operation that produced an annual flow of 150–200 kits.

When I first met Paul in 2002, he was single and lived alone in Sybille Canyon with the ferrets while the technicians that worked with him commuted the two-hour round trip up from Laramie each day. Upon first shaking his hand, I noticed that he had the sharp personality and watchful eye needed to ensure the conditions and care for such a sensitive animal. Originally from Philadelphia, Paul completed his master of science degree from the University of Wyoming by studying ferret behavior in South Dakota and evaluating the detectability of ferrets via night spotlighting surveys. Like a field biologist, to succeed at his job of overseeing captive breeding of black-footed ferrets, Paul dedicated and set the rhythms of his life to ferrets. He lived in a small home on the captive breeding site because caring for ferrets in captivity was not a 9-to-5 job. It required continuous attention and ability to respond to emergencies at all hours of the day and night. Ferrets had to be monitored and fed, and their cages had to be cleaned daily throughout the year. In springtime, monitoring of females was required to determine when they were in estrus to pair them with mates. Then, forty-five days later, they had to be monitored for births and the status of litters. In the summer, there was a need for monitoring and caretaking of vulnerable kits and their mothers. Then, in late summer and fall, kits were preconditioned and transported for release on reintroduction sites from Mexico to northern Montana. After the kits were shipped out to sites by October or November, planning for next year commenced as pedigrees were assessed and potential mate pairings were mapped out for February and March of the coming year based on a formula that aimed to maintain the greatest possible genetic diversity of the captive population. There was little time for vacation, and even less time for a personal life.

. . .

To rear the rarest mammal in the world required specially designed equipment. In Malta we built plywood nest boxes to Paul's specifications for female birthing. A nest box was a small, sturdy, two-chambered box with a four-inch entry door from the top that could be locked with a latch. At the bottom of the box, a hole was drilled and four-inch-diameter black corrugated tubing was tightly connected that led to a larger plywood box with a Plexiglas front and screen top to resemble above ground exposure. All boxes were kept sterile and painted bright white. Outside, sixteen adjacent pens, each thirty-two feet long by thirty-two feet wide, were erected directly into and onto the prairie. Prairie dogs captured on local cattle ranches were brought in by the hundreds, quarantined, killed, gutted, organized into individual plastic bags, and frozen for a year's worth of ferret food. Valerie had already established a colony of more than one hundred hamsters housed in metal bins in a nearby garage bay that would be used to feed young kits and train them to kill live animals. This was a gentle introduction to predation, as the small puffball hamsters were less likely to hurt young naïve kits than would the larger prairie dogs with strong jaws and razor-sharp teeth.

The delicate cycle of captive breeding began in earnest in March, when we became novice reproductive physiologists. Pairing normally solitary male and female ferrets together for extended periods of time can be dangerous. Long canines and muscles built for taking down prey nearly twice their size when used on each other can result in severe injury or even death. Thus, we carefully monitored the precise timing of when individual females entered estrus, using pipettes of water and microscope slides to perform vaginal washes. Counts of more than 90 percent of cells being keratinized indicated that a female should be ready to be bred successfully. We then placed a sexually active male with the female in an outdoor nest box for three days, hoping they didn't instantly fight and listening for a struggle in the plywood box as we shut the door.

Following a pairing, we rested the male for three days prior to pairing him again with another estrus female—allowing his seed to restock. We similarly let the female rest, conducting an additional vaginal wash seven days after the initial pairing to determine whether ovulation had occurred. Given that all black-footed ferrets give birth around six weeks post-conception, using this initial pairing date we were able to fix the time when the females were likely to give birth or whelp.

Using this recipe, by June 2002 we produced eight litters totaling thirty-four kits. Thirty-four pinky-finger-sized young with their eyes closed wiggled in a pile of thin fur and pink skin. We left them alone in their indoor whelping boxes with lids closed tight. Just as we were nervous in pairing males and females, we were nervous of the mother rejecting or killing her young. To avoid tipping them into an infanticidal killing frenzy, we kept disturbance to a minimum; at first we opened a nest box lid only to change the bedding every five days. Despite this sensitivity, by July we had lost three kits. Two of the kits, when found, had been largely cannibalized by their mother. Was it a natural death of the kit and the mother simply ate the available carrion, an instinct to take what she could get to continue producing milk for the others? Or was it confusion by the mother, and infanticide as we had feared? Elsewhere in the animal kingdom the non-nurturing urge to kill off another mother's kits to limit competition is fairly common. In this strange captive setting, did she not realize the kits were her own? Whatever the reason, with so few litters to raise, we hoped it was an isolated event. In another nest box, where one kit appeared to have died from an airway blocked by eating a small square of cardboard bedding material, we blamed ourselves.

. . .

By August, when the kits were well along toward feeding for themselves, Valerie and I began to alternate weekends off and I visited my parents, who had recently relocated for work at the newly created U.S. Geological Survey Northern Rocky Mountain Science Center in Bozeman. My father and I took my mother out to dinner for her birthday. Bad Thai food at a strip mall still tasted good because it was the first restaurant meal I had in a long time. I was at home, but I still felt restless without the ferrets to care for. They had become my crutch, my reason for leaving J, the only justification for a life away from a woman

FIGURE 5. Prairie dogs captured in live traps for transport.

who loved me. I needed something, anything to give me a home, a peace of mind that reaffirmed that I was living a useful life and that I was not just passing through a cycle of travel trailers and temporary jobs. I needed something to give me traction when all idle thoughts seemed to slide toward loneliness. In an act of desperation, I went to the dog pound and adopted a red heeler, named her Abby, bought a sack of dog food, and headed back north to the Hi-Line.

When I got back to Malta on a Sunday evening, Valerie had left me a message that an entire litter of six kits died over the weekend. I wondered how, why, and if somehow I was to blame for taking two days off. We shipped the bodies down to the Wyoming State Veterinary Lab in Laramie, uncertain of what went wrong and anxious for the results. We worried that other litters could similarly collapse, so we once again cleaned the cages and disinfected the building, only with a new vigor. In the end, however, we could only leave the mothers to rest and care for their young.

Resigned to the will of the ferret mothers, we knew we could not disturb them with more attention, so we buried ourselves in the work of trapping prairie dogs for ferret food north of Malta. Saving the landowner the time and money of poisoning prairie dogs, in a single week we collected 516 individuals and placed them in elevated cages in an old tin shed with a concrete floor for a two-week quarantine. That was the set period of time during which a disease such as sylvatic plague would have run its course, and if they survived we could reasonably declare them disease-free ferret food.

While the prairie dogs were in quarantine, we spent the following weeks preparing the outdoor pens for ferrets. We checked for breaks in the first or second level of our fences that kept the ferrets in and predators out, and measured the voltage and cleared grass away from the electric fence wires strung to keep raccoons, skunks, badgers, and any other ground predators from climbing into our compound. Within each pen, we blew smoke from smoke bombs down into burrow systems with a leaf blower to check for escape burrows and filled in any breaches with gravel or concrete. We trimmed the grass down to the ground to resemble the closely cropped vegetation of a wild prairie dog town in the dry, clear days of midsummer. Last, we scrubbed out underground nest boxes that we hoped ferrets would use to spend the daylight hours and thus allow us to easily trap them for checkups. The nest boxes also made it easier for us to clean out their feces to prevent bacteria buildup and infections and to remove the accumulation of prairie dog fragments that attracted blow flies. These small flies were the bane of animal caretakers because on any scratch or wound they would lay eggs that would turn into flesh-eating maggots that burrowed just under the skin of our little brood of multi-thousand-dollar animals.

Once the prairie dogs passed quarantine, we selected a lucky few to pretend again to be wild and free. We released two or three into each cleaned-out pen to excavate the old burrow systems, performing prison labor outdoors after two weeks in a suspended square metal cage. Previously, the only boundaries for these animals were social divides between families on flat prairie. A lifetime of only imagined boundaries was replaced by real boundaries and confinement with strangers. I watched the selected few prairie dogs quickly readjust to the view of the sky, taste of fresh grass, and feel of dirt between teeth and toes. I silently urged them to enjoy their time in this semi-captive state, perhaps even escape, wishing I could relate to them that they had only a few days before they were to be recaptured and soon after butchered and put to use as ferret food along with the rest of their captive cohort.

While the captive prairie dogs were declining toward a feeble state on a diet of dried grain and alfalfa pellets, the ferret kits were getting feisty and outgrowing their indoor plywood boxes. Despite their size, which nearly equaled their mothers' so that the nest box was crammed with writhing bodies, we stuck with Paul's strategy of keeping families intact and indoors until fifty or sixty days after birth. We defrosted prairie dogs from our line of large chest freezers, quartered their bodies, and fed the torsos to the females and kits that devoured the tender, marble-sized dark red organs inside. I left the tougher hind legs for adult males that sat alone in their pens, useless until the next breeding season.

We checked on the males, but we did not dote on them like we did the families. Captive breeding reduces species not so much to individuals but to numbers. Studbook serial numbers correspond to each ferret born since the final wild ferrets were captured from Meeteetse, reflecting each individual ferret's genetic lineage. Lineages are made to mismatch so that genetic diversity is maximized, to the extent it can be from an origin of at most eight genetic founders. Biology overtakes sentimentality, because despite their inherent value, males can breed repeatedly with multiple females and have no role in litter rearing. It is all up to females to produce a crop of kits each year. We lost sleep only for the females, worrying whether they would be good mothers, checking to see whether they were indeed good mothers, and hoping they could raise their three to five ferret kits to adulthood to justify our work.

By the end of the week, Valerie had gotten the lab results that told us we likely lost the litter of six to coccidiosis, a bacterial outbreak. We knew there was nothing more we could have done, but that didn't satisfy us. A single death when our output was already less than forty affected years of work and planning. Perhaps we could have disinfected the cage more often, but we wanted to avoid disturbing the mothers and their kits more than necessary. Perhaps their inability to avoid deadly bacterial infections was a consequence of inbreeding and the narrow genetic bottleneck, a compromised immune system with diminished ability to fight off infection. Nobody could really know what was going on in the instinctual processes of these secretive small mammals, and now we also had to concede that we didn't know to what degree inbreeding might have harmed their physiology. Our only fallback was to stick with Paul's strategy of what worked above all else.

With the critical period in ferret litter survival at hand, I skipped my next free weekend and stayed near the trailer and ferrets, checking on them twice a day but otherwise leaving them in peace. I took time to read, cook a regular meal, do laundry. I tried to call J to talk for the first time in weeks, but she was away, on a fishing boat off the coast of

Maine. I left a message, asking her to call me back, listening to her answering machine so that I could at least hear her voice.

. . .

August 30, my birthday, and a letter from J. The paper was dark grey, handmade in Kathmandu, pen and ink drawn on the front and intricately cut and folded to make into a street scene:

> Thinking man, who when I crave, will wait for me, who when I cry, will make me laugh. Point to my heart and belly and ask what is at the end of this rainbow? You muddle my mind, I welcome your expanse.
>
> May your days be colored with fire. Life, commit to growth, wink for love, feel your import, your beauty . . . keep smiling! Have a Happy Birthday Love

I celebrated my life with death. The prairie dog quarantine period was over, and it was time for them to restock our row of chest freezers. An assembly line–style butchering of prairie dogs we termed "processing" in a veiled attempt to downplay the emotion of taking hundreds of lives in an afternoon. In the days leading up to the event, I was disgusted with myself for killing the declining prey of an endangered species. In the coming years, multiple environmental groups would petition for the listing of the black-tailed prairie dog as a threatened species. But for the sake of captive black-footed ferrets, biologists and refuge employees from across the area drove in from up to three hours away to help. Some even volunteered to join in, unfazed by the task, likely by virtue of the prairie dog still being considered by many as a nuisance animal.

I took the lead on handling the live animals and transferring them into and out of the gas chamber. I hated the job, but I did not want someone who was reluctant to kill and was hesitant in acting, allowing the prairie dogs to awaken from the anesthesia before death. I also didn't want someone so eager and bloodthirsty that they did not allow the animals to go completely to sleep before humanely dispatching each one quickly with the snap of its neck. I selected a dozen animals at a time, placing the cage in our makeshift plywood gas chamber. I closed the lid and turned on the carbon dioxide tank, waiting for the animals to black out, lose control. Once on their sides, I opened the plywood box and quickly dislocated their necks one by one, passing their limp bodies on to a table where others sat to eviscerate, place the cadavers into individual Ziploc bags, and label with the processing date.

After the first batch of a dozen prairie dogs, I went numb, repeating the motions without thinking. In moments of hesitation I tried to focus on the ferret families just down the road. When that wasn't enough, I focused on my resolve for the project, pushing away self-doubt so as not to show it in front of peers and superiors whom I respected and from whom I wanted respect. I thought back to my time working with my uncle on a ranch on the south fork of the Shoshone River near Cody, Wyoming, of pushing the cattle up into the forest for the summer, but down in the valley still needing to care for a single cow with a genetic defect that caused her hooves to continue growing to the point of becoming like large shoes. The cow, appropriately named Goofy Feet, fumbled along to keep up with the other cows. I remembered making my first trip to the slaughterhouse in town, watching them unload Goofy Feet from the trailer and seeing her smell death in the corral next to the processing room. I saw her mind buck generations of domestication to register that primitive threat of death when coming upon a place where a predator has already struck. With her *Bos taurus* Asian stock origin, I imagined her registering an image of coming upon a tiger kill, fearing the large cat is still nearby and ready to feed again.

I watched Goofy Feet bolt for freedom through the narrow gap between the trailer and the gate. I chased her down out of instinct and order, keeping her from running down the hill and into town, and herded her toward the others who were driving up with lassos. Once she was tied to the bumper of a Jeep and hauled back toward the killing room, I regretted my part in her recapture. I remember going into the killing room ahead of her and seeing the men go back to work on the next animal, a black Angus bull. The special silver gun with a metal piston was placed to its forehead, shooting a captive bolt through the top of its skull and fatally stunning the animal. Quickly hanging the animal by its hind legs up into the air, the men placed a large metal bucket underneath and cut its jugular to spill out the life blood.

Since then I had killed goats, chickens, all for food for the table. In the Philippines I even helped kill dogs and house cats out of conservation need, but I couldn't get used to the necessity of killing prairie dogs. It seemed counterintuitive—killing the very animals ferrets relied on to recover, so that we could condition young captive ferrets to the taste of prairie dog meat.

By the late afternoon I struggled to keep up with wringing necks prior to when the final animal in the bottom of the chamber began to wake. We finished the last prairie dog for the day, number 401, and I reached down for a bottle of water, but the ringing of thick necks had taken a physical toll—my wrists were sore and hands no longer able to grip. My leather gloves were wet with sweat and the urine of prairie dogs that peed on one another while succumbing in the gas chamber.

After the makeshift crew went home for the day, I retreated back to my trailer behind the mechanic's shop and opened up the bottle of Old Crow whiskey stored under the sink and a can of warm beer. I really wanted a cigarette but didn't have tobacco, nor the resolve to drive into town. Over the entire summer, we would process more than 1,685 prairie dogs from the Great Plains of Phillips County for our freezers. My only hope was that the math would work out for us and we could justify this carnage with a decent number of ferret kits produced from our facility. I turned on the TV and got no channels, put in a movie but could not focus, sitting on the foldout couch and staring at the paisley squiggles on the wallpaper just beyond.

. . .

The following week, we released the five surviving litters into the outdoor preconditioning pens. Tender black feet and neat little masks, the nearly full-grown kits were feisty, quick to hiss and express their scent glands. They poked their heads out of the corrugated black plastic tubes and darted back into their nest boxes. We captured them one at a time and then their mother, placing them together in a single outdoor pen apart from other family units and males. A first step toward freedom, preparing them for the scents of the natural world away from their sterile, air-conditioned existence and into the heat and dirt of the prairie.

There was an odd sense of relief, and at the same time resignation, because there was nothing left to do but put out food and hope the kits survived. Our only way to check on them was whether they used the nest box inside the pen or came out for food. They could easily just hide in a burrow and die or try to escape, yet to do so would require them to burrow under cement walls sunk seven feet into the ground. Even if they somehow did burrow deep enough, we had a second perimeter fence with live traps in the corners that we checked daily for escapees. That outer fence was the last barrier before they would get to a harsh outside world, far north of suitable habitat for ferrets and not a prairie dog in sight.

The evening after their release to the pens, I raised a celebratory toast to the setting sun. I sat in my trailer hoping the ferret kits felt safe outdoors, that they would venture outside of the nest box and burrows under the cover of darkness to smell the night air, feel the coarse blades of dry grass on their bellies, dig into the surface of the earth. I hoped that their instincts reconnected them with the outside world. So much relies on that connection with the wild. I reflected on this job in creating and raising sensitive lives. I thought of the role captive breeders of rare species play and how we dictated wildness in these animals' lives. For the prairie dogs to enter captivity prior to slaughter must be crushing, and I hoped the release of ferret kits to their first taste of nature outside of the laboratory trailer would have the opposite effect. I wondered whether the kits and their mothers now, at last, felt safe in the environment for which they as a species had spent tens of thousands of years developing such a specialization. Whether they would be able to learn and interact as a family in this stepping stone to the harsh, unrelenting truths of life in the wild, relying on the instinct that calls them to escape from our protection but at the same time stay in the company of their siblings.

As humans, we have a similar innate desire for freedom. As Ed Abbey says, "Wilderness is not a luxury but a necessity of the human spirit." Yet during long hikes alone in the woods or time secluded in a cabin, the mind of even the most stalwart recluse drifts to memories and thirst for human conversation, human touch. We need that time in the wild, yet we are fundamentally social animals. I thought back to the last time I felt that visceral contrast, during a typhoon in the Philippines with J, reluctantly waiting for her late into the evening at a local rum bar, far away from my thatch hut, playing cards with the shop owner while J was down the street editing an instructional video on the harm of dynamite fishing to coral reefs. I remembered how the rain came down in sheets outside under the flickering streets lights of a third-world power grid. How the bar was vacant except for Edmund, the owner, and his plump high school daughter. How Edmund's daughter stood behind the bar making eyes at me as we played cards and drank rum into the evening. Boards were left down on the wide street entrance to his bar to let in the waves of tropical heat and humidity from the storm. The bar's yellow lights shone dimly on a handmade plywood table covered with a greasy plastic sheet, surrounded by plywood stools on a concrete floor. An old TV that was always on yet silent in the corner showed lines of static over what looked to be an old Chuck Norris movie.

The images of smoking buildings out of the corner of my eye only struck me when the newscaster stood in front of the scene. I felt like I was coming out of a dream, wondering whether the images on TV were real or whether it was a movie. Edmund noticed me watching and had his daughter turn up the volume. There in the static were the two towers of New York burning and crumbling, with the broken voice of a newscaster speaking Tagalog and telling the world something was happening. I had not been out of the jungle in weeks, and I asked Edmund if this happened today or a week ago. He told me it was "today, just now," and we listened to words like *terrorists* and *air attack*.

The typhoon was escalating and the power went out. Darkness, and I suddenly wanted to be somewhere, anywhere, just not there trapped in a small bar in a rain storm on the edge of civilization. I went out into the rain, found a motorcycle taxi, and rather than retreating to my sanctuary in the interior of the island, I went to the editing studio where J was working. When I arrived, she was standing just inside the doorway, asking me if I heard the news. My instinct was to go back to my hut in the jungle and hide farther away from it all. Powerful tropical winds and the crush of large waves echoed up the city block from the coastline. Sounds that at one time I remembered to be comforting now seemed chaotic, looming just beyond in the darkness. Yet when I asked her to come, J would not concede to the wet half-hour motorcycle ride inland, instead talking me into renting a room in town.

We undressed from our wet clothes under the light of a candle in the small cement room, painted white and only large enough for a bed and small school chair. A sink and toilet were squeezed into what must have at one time been a small closet in the corner. The steel bars on the window and thick cement walls of what otherwise would have felt like a prison tonight felt like a form of protection. The power outage had left the city black, and we lay on top of the damp sheets as the rain poured down loudly on the tin roof in thousands of hammer strikes before flowing down to flood the streets and sidewalks below.

"Are we at war?" J asked me in a soft voice as I blew out the candle the front desk had given us.

"I don't know," I told her. "Maybe."

. . .

I hated phone calls, the detached contact with another human, the denial of the primitive need for all primates to be able to visually judge the joy, remorse, or sarcasm on another individual's face. But I especially hated phone calls to Randy, stationed more than a hundred miles away, reporting our unexplained loss of ferrets in the pens.

"We lost one litter of five."

"Crap."

"We found two of the carcasses. Lab results say they likely died of a coccidia outbreak."

Randy was silent, and before he could reply I got the rest of the bad news out.

"We also lost one litter of two and one litter of three. They ate when we fed them until a few days ago and then we just never saw them again. We set traps but they didn't come to the bait. Likely they're dead in the burrows deep underground."

"Not good."

Randy told us to not get down despite such substantial losses. He told us to focus our energy on the remaining kits and keep extra attention on them. But it was hard not to question every move and every chance we might have had to capture the kits and bring them back into the protective shelter of our laboratory trailer. I wondered what Randy really felt, whether he passed on the information to others or was too embarrassed to give higher-ups the play-by-play. I wondered whether our jobs were secure.

Walking back to my trailer at the end of the day I was told J had left me a phone message. I remembered that she had called the refuge headquarters a few other times trying to reach me over the previous few weeks. I just never had the energy to call her back. Work held back the depression until I got back to my trailer at nightfall, then feelings of self-doubt crept in. I brooded on the realization that we were failing, that we were losing far too many kits to justify our work. I told myself that I wanted to call J, but I used the two-hour time difference as an excuse at the end of the day. My internal conversation went something like, "Well, she would likely be asleep by now or she is likely out with friends and doesn't have time to talk." But tonight I decided to make the call from the maintenance building. After a quick hello, I noticed her voice was different, halting, somber. I regretted calling; there was silence and she quickly realized that I knew she was being different and got to the point.

"I think we should let this go," she said.

"Really? Why?"

"I don't think we are on the same path. We are different people."

"I'm the same," I asserted.

Silence.

"Are you the same?"

Silence.

"Is there someone else?"

"It is not like that. It is just too long. Too far to keep this going."

"So you haven't found someone else?"

"No, but I do have feelings for other men."

"Love?"

"I don't know, just feelings."

I tried to argue, but had nothing left.

. . .

The remaining kits were strong, healthy, and bigger than their mothers. They began to come above ground alone and take chunks of prairie dog back to their individual burrows. We listened for the squeaks of excitement as the food was brought back to a burrow by the brave sibling and shared with brothers or sisters. It was nearly time for release into the wild, and Randy came up from Lewistown with the anesthesia trailer and equipment. We captured the six remaining kits in wire traps baited with pieces of prairie dog. They were aggressive and nipped at our hands as we reached down to move the traps to the trailer. Once inside the trailer, we put on surgical masks and gloves. Randy took the lead on the first kit, coaxing it from the trap to the dark, homemade anesthesia tube made of white plumbers' PVC plastic, inviting in its darkness like a burrow. He sealed it in and hooked up the hose that brought in carefully calibrated amounts of oxygen and isoflurane gas. In silence, we prepared the syringes and vials as he checked the animal to judge when it was under sedation. When the ferret could no longer right itself in the tube, he pulled off the PVC pipe ends and poured the long ropelike animal onto a clean white towel that was draped across the operating table. He placed a small mask over its muzzle that fed in small amounts of isoflurane gas to keep it just sedated. After attaching a pulse oximeter to its hind leg to monitor the kit's heart rate, he checked its teeth, which were complete with a full set of sharp, quarter-inch-long, gleaming white canines. These were the essential tools for killing and gripping. When a ferret reaches the age of four years, at the normal end of its brief life span, it is not uncommon to see its canines yellowed and split lengthwise or sheared off at the tip or base. Randy checked the ferret's sex and body condition, looking for any physical imperfections that could harm its chances of surviving the next step of release into the wild. We recorded every detail as he prepared the injection areas with alcohol and Betadine, turning the intricately layered buff fur with thin black guard hairs to a temporary auburn brown over the shoulder blades and pelvis.

We checked the PIT tags with a battery-powered reader to make sure they transmitted their unique nine-digit identification numbers. These microchips, the diameter of a grain of rice, were implanted just under the skin of the ferret and relayed its identification numbers when scanned by an electronic reader. Each ferret had unique numbers in two tags, one above its hips and another above its shoulders. The numbers could be traced back through data sheets to the time the ferret was first implanted with its microchips, where it came from, its sex and previous body condition. All of these were key bits of information when an animal is later found in the field, allowing biologists to identify individual ferrets by placing electronic readers by burrows, minimizing the need to trap and handle the animal. They allowed us to piece together where the animal had been, how it is doing, and, based on those patterns, we could begin to guess what might happen in the future.

Randy inserted the large hypodermic needle just under the skin, placing the microchip tags in the fore and aft of the ferret and then quickly sealing up the two small wounds with surgical glue. After he drew blood for medical tests, he gave each kit an injection of canine distemper vaccine and a dose of penicillin to help it on its way. Within five minutes the entire procedure was complete, and Valerie and I took turns processing the remaining ferrets for release.

After finishing the last ferret, we stepped outside, leaning our backs against the trailer as we breathed in the fresh air and took off our gloves and masks. Randy explained that higher-ups miles away had decided that the six kits we produced were to be released the following day on a reintroduction site in southern Phillips County. Further, Randy and the larger ferret program had decided not to stock our Malta facility for the coming winter, and to focus instead on preconditioning ferrets reared by Paul at the Sybille facility next fall. We knew what this meant. It meant that we would not be breeding females with males, watching births, and raising kits. The national program thought our production rate was not up to par with other facilities—the math didn't work out and it decided we were a net drain on ferrets to the program.

. . .

The next day I drove a pickup truck with the kits in travel cages due south from Malta to a cluster of prairie dog colonies near the Missouri River. The two-hour drive on dirt roads gave me time to think. I questioned why I chose this path. Why I was in the center of

Montana? Why I chose a state over a good woman who I thought, perhaps someday, I would marry. Why I denied what I should have known—that it would end badly like this, with nothing but open land and my thoughts for comfort while she moved on with another man.

I hit the Dry Fork Road and headed due west past the Veseth Ranch and then the Jacobs Ranch, where they were busy preparing for the coming winter by putting up hay and cleaning out corrals just before the cattle came in from their summer range. This business of restoring ferrets felt so difficult. Why were these wild animals so hard to bring back? I thought back to when I was a teenage boy working on Hardpan Ranch just outside Yellowstone National Park, and when my uncle took me to a steak lunch hosted by the federal government to tell area ranchers of the proposed wolf reintroduction to Yellowstone. The federal biologists knew that everyone in the valley was against it, and I remembered feeling sorry for biologist Ed Bangs as he sheepishly walked over to our table. He was burdened with the knowledge that we knew the powers above had put him in a hard position, having already decided that there should be a release of wolves brought down from Canada, leaving only the questions of when it would happen and how they would be controlled when they left the park. Ed knew that we knew we were never really going to get a chance to have input, that it was all taken care of and the steak lunches were just for show. It had been an event to make the newspaper readers in town and back East think that the government cared about what people making a living on the ground thought, and that they weren't just telling us how it was going to be.

But I now knew that the difficulties of wolf reintroduction paled in comparison to conservation of the black-footed ferret. Boiled down to its most incremental part, we had failed even to make more ferrets in a controlled, test tube–like environment. Like a slap in the face, Malta had taught me that conservation was infinitely more delicate and difficult for ferrets than for wolves or grizzlies. Our three-year experiment, in which 132 ferrets were held in captivity at the Montana facility, resulted in forty-six mortalities and only produced six new individuals.

I released the six kits on prairie dog colonies within high, open prairie land managed by the Bureau of Land Management. There, just a mile off the Dry Fork Road, a dense cluster of prairie dog colonies formed the core of the Montana ferret reintroduction sites in southern Phillips County, just above the Missouri River Breaks. The following week I took the nine remaining adult ferrets still in Malta down to Paul in Sybille. Our experiment with a Montana captive breeding program was over. In the coming years, nearly all the captive breeding programs in other states would similarly be decommissioned, focusing attention and responsibility on Paul. The burden would all fall on his shoulders once again.

CHAPTER 5

Fall

It takes eight minutes for sunlight to reach the Earth, for a photon in a wave of light to travel 93 million miles and bounce off this planet. Distances in central Montana can be similarly hard to fathom, horizons measured by flat views so clear that novices always underestimate the time and distance between landmarks. Stark waves of land where, unlike the railroad town of Malta, there are no grain bins or silos in sight. Just three-strand barbed wire fence on old cedar posts and open range. Homesteads tucked into draws out of the wind, separated by intervals that would require different zip codes on the East Coast.

You take the gravel road due east from Highway 191 for an hour until reaching a T-junction with Dry Fork Road at the Second Creek schoolhouse. Turn south down a two-track road past the last homestead, where Judy Blunt lived for a time as a ranch wife before uprooting and heading to Missoula, where she wrote a memoir of her escape from remote domestication entitled *Breaking Clean.* You cross into Missouri Breaks country, broken topography that hides a river that drains much of Montana during the spring thaw and lies cool and slack, hidden under the last bench on the distant horizon. Ahead, the land rises and fractures along 125 protected, unpopulated river miles.

Other ferret reintroduction sites occur near seldom-seen towns named Interior, Cactus Flats, Seligman, Dinosaur, Medicine Bow, Raton, and Janos. But UL Bend is the most remote by far, requiring a two-hour drive from the end of the blacktop. The UL Bend is a distinct,

FIGURE 6. Upper Missouri River near the UL Bend National Wildlife Refuge, Montana.

epiglottis-shaped dip of the Missouri River in the heart of the million-acre Charles M. Russell National Wildlife Refuge. It was first described by Lewis and Clark on May 21, 1805:

> The Missouri in its course downward makes a suddon and extensive bend toward the south, to receive the Muscle shell river, the point of country thus formed tho' high is still much lower than that surrounding it, thus forming a valley of wavey country which extends itself for a great distance in a Northerly direction; the soil is fertile, produces a fine turf of low grass and some herbs, also immence quantities of the Prickley pear, without a stick of timber of any description.

This bend in the river has been a key north–south migratory crossing for bison, elk, deer, and antelope en route to southern and more mild range conditions during the winter. A siphon of land that collected migrants reluctant to cross the flowing waters of the Missouri. The abundance of game attracted Assiniboine, Gros Ventre, and Blackfoot tribes during the large-scale seasonal movements of wildlife through the area.

Although bison have been extirpated and with them knowledge of their historic migration pattern, pronghorn still make an epic migratory passage through UL Bend from their summer range in southern Saskatchewan to their winter range in central Montana, just north of the town of Roundup. This three-hundred-mile round trip is made more difficult by the fact that antelope evolved without fences. An animal built for speed that has outlived its evolutionary counterpart, the now-extinct American cheetah, the antelope has yet to figure a way over countless miles of barbed wire marking pasture and property lines. Sometimes pacing along fence lines for miles until they can find a gap where the lowest strand of wire is raised far enough from the ground for them to ease under. Those pronghorn that stay north risk blizzards and trains as they search for high ground on the elevated tracks of the Great Northern Railway, getting plowed down by locomotives running east and then west each day. Spring thaws reveal piles of duff brown pelts and black pronged horns in railway ditches. Those that head south past UL Bend must swim across the Missouri River, a formerly narrow gap that now is flooded by the Fort Peck dam. Some winters, they drown in droves if the ice is thin or there is a record spring flood. Obstacles abound.

Joel Berger and other prominent ecologists warn that long-distance migrations, or LDMs as they call them for short, should be among the most urgent conservation priorities of our generation. Increasingly, LDMs of large herbivores have garnered public interest for a number of high-profile species, including wildebeest of the Kalahari and chiru of the Tibetan plateau. Here in one of the richest and most developed countries in the world, new research suggests that the age-old migration of pronghorn using UL Bend rivals the more famous migrations of these other animals in both distance and endurance—an extreme example of wildlife being wild. However, because of their large geographic scope, Joel and others are finding that LDMs are highly sensitive to habitat conversion and construction of roads, railways, and fences—the very same anthropogenic threats that eventually can lead to extinction of entire species. Thus, first goes the wild behavior, then the wildlife.

. . .

Driving down to UL Bend there was a hundred-mile cumulus view. I dipped down into the valley just above the Missouri River and rode the bench eastward in my tired, government-issue truck. My eyes strained to change focus from a fixed landscape view to one just ahead of my front bumper as prairie dogs scurried to their burrows in fear of the first vehicle they had seen in weeks. Their large rodent societies lived on the flats above the river's edge, visible life after miles of dried-out yellow grass stalks and open plains. Females and males from family groups left small clouds of dust as they shot away from the stubble grasses toward

FIGURE 7. Black-tailed prairie dog *(Cynomys ludovicianus)* colony in the center of UL Bend National Wildlife Refuge, Montana.

specific, raised burrow entrances, diving down burrows with force and flickering black-tipped tails.

These black-tailed prairie dogs would not sleep through the coming winter, entering into a state of torpor and waiting for spring green-up like other prairie dog species to the south and west. They are active year-round, trimming grasses and forbs short to the ground. They girdle noxious sagebrush that blocks their view of predators, so that the only things shooting above prairie dog eye level are the cottonwood trees growing from the uninhabited draws that flood in the spring. These flat lands with open ground and good sight lines are also valuable nesting habitat for horned larks, killdeer, and the increasingly rare mountain plover, an inappropriately named migrant that typically nests in the middle of prairie dog towns.

Prairie dog family groups, or coteries, typically adjoin one another, creating large, open areas with seemingly continuous burrow openings and cropped vegetation known as colonies. Bison and pronghorn, the hallmarks of the Great Plains, both prefer to spend time on prairie dog colonies. For feeding, bison tend to select young colonies or the edges of older colonies where the vegetation is dominated by nutritious

FIGURE 8. Ferret Camp.

graminoids and spring forbs. For resting and wallowing, however, they select the more open centers of colonies that are dominated by forbs and dwarf shrubs. Lanky pronghorn similarly prefer the open centers of prairie dog colonies. This seemingly counterintuitive use by multiple herbivores of common areas where grasses are cropped short is likely because of nutrition, where, like a suburban lawn, grasses that are kept cropped short do not easily go to seed. Thus, they retain nutrients not present in other areas where grasses are fibrous and senesce after reproduction. Studies have shown that bison feeding on vegetation in prairie dog colonies typically consume less dead plant material, and what they consume is often higher in protein content and more digestible than what they obtain away from prairie dog colonies.

Yet it is undeniable that prairie dogs consume grasses and reduce the overall amount of forage biomass available to ungulates. Thus, there is a trade-off between quality and quantity that prairie dog colonies offer to large herbivores, and it is this tradeoff that perplexes many domestic cattle ranchers. Upon first looking at a close-cropped prairie dog colony, anyone can notice that there is obviously less food biomass. But unlike bison and antelope, prairie dogs cannot get up and move when a drought occurs or the vegetation is consumed to a low level too early in the summer. It takes careful management to care for the range when trying to maintain prairie dogs while raising sedentary and fenced domestic cattle. Because when prairie dogs are combined with fenced, domestic cattle for extended periods of time, any benefits of increased nutrient content for weight gain of cattle can be counteracted by the need of both species to feed their fermentation vats. Sharing the available forage may be suitable for both herbivores in wet years, when studies have shown that prairie dogs prefer to settle in areas heavily grazed by cattle and other herbivores. But in dry years, the combined effect of prairie dogs and larger herbivores can reduce vegetation to its roots, turning prairie to bare dirt. Thus, to keep prairie dogs on many western lands, managers must have the foresight both to pull cattle off prairie dog colonies during dry years and to put cattle into pastures with prairie dogs during wet years to maintain a balance between both species. This dynamic management, in part, mimics the past interaction of bison and prairie dogs. Except in place of movement and migration, in the now barbed-wire–bounded world of the Great Plains, managers must actively manage herbivores in response to the cycles of rainfall and drought that shaped the plains over tens of thousands of years and continues today.

Below ground, prairie dogs are master engineers, pushing and digging in the prairie earth. They are called "ecosystem engineers" by ecologists, a label also given to kelp in the productive, coldwater coastal zones and to beavers with their dam building and associated flooded pools. Prairie dogs aerate the soil, facilitating the penetration and retention of moisture, exposing nutrients in soil to the surface, and providing miles of underground three- to four-inch-diameter burrow systems used by a host of invertebrate and vertebrate species. Burrow openings often are occupied by secretive black widow spiders, and moist burrow depths are used by six-inch tiger salamanders to survive through the eleven months of the year when the surface of the prairie is inhospitable to them. There are also burrowing owls, voles, mice, rattlesnakes, bull snakes, swift foxes, badgers, and other conspicuous and cryptic species that use prairie dog burrows for cover.

The combined effect of prairie dog activities on the landscape leads ecologists also to label them a "keystone species," meaning they are a species that has unique, significant effects on the ecosystem out of proportion to its abundance. Prairie dogs are members of this relatively exclusive grouping of species because in addition to their colonies hosting more than one hundred other species, prairie dogs themselves provide essential seasonal food for many terrestrial predators and migratory ferruginous hawks, golden eagles, and other raptors. Perhaps most notably, prairie dogs are unique because the black-footed ferret occurs only on their colonies.

. . .

The bend in the Missouri River where Ferret Camp resides did not receive its full modern name until 1894, when Oren and Will Bachues established a ranch and marked their cattle with a distinctive "UL" brand. After thirteen tough years, distance, weather, and poor soils forced them to give up on this bend in the river and move north to the town of Malta, where they set up business as dairy men and then hardware merchants. Oren eventually moved farther north from UL Bend to the big city of Calgary, and finally settled down in Seattle, where he died at the age of eighty-eight.

Today, similar to the Bachueses, all of the ranches and homesteaders near UL Bend have gone bankrupt or have been bought out, leaving a broad swath of undeveloped government land. Ferret Camp is the only sign of recent human habitation on UL Bend: a neat east–west row of eight camper trailers that are retired to weather away as field housing at the end of the road in Phillips County, Montana. The camp got its name from being the first place in the state, and one of the first places in the country, where biologists reintroduced black-footed ferrets. There are no power lines; Ferret Camp runs on propane and car batteries charged by solar panels. To live there, you pack in water and food. The seclusion is broken only by occasional pulses of summer life as college researchers come to study the relatively intact fauna of the mixed-grass prairie in the area.

I settled into the westernmost trailer that had a "4" painted on the outside. All the other trailers were empty and available, but this one offered the best view and contained a full set of dishware in the cupboards. Outside the trailer door, prairie dog families came above ground. The camp was set up on the north edge of one of the largest prairie dog colonies in the state of Montana. The four-hundred-acre Locke prairie dog colony (named after the Locke homestead that used to be here) still contained a few remnants of past human habitation. A six-foot-tall hedgerow ran along the slope uphill of a squat, rectangular barn made of stacked, massive cottonwood timbers. No chinking between the logs, the barn sat more like a fort against the wind than a full shelter from the elements. It had a dirt floor and tar-papered roof that had been blown off and replaced multiple times, most recently with plywood and hastily nailed-down shingles. The barn was now used as

FIGURE 9. Young black-tailed prairie dog collecting grass for its burrow.

storage for everything that could be blown away in a powerful gust or decay in the sun and rain: sage grouse nets, oil and gas cans, old car batteries, four-wheelers, and regrettably, an old telephone. That contraption connected the outside world to camp by a thin line of wire buried underground for miles to reach this point. A spider web of human connectivity that had one outward strand that miraculously emerged from the ground just inside the twelve-foot-wide barn doorway. The old phone sat silent most of the time, only working on days when there was no rain and the barn rabbits hadn't chewed through the cord.

Apart from the barn, everything else was temporary. The neat row of trailers was tucked uphill of the barn but just inside the east–west hedgerow. An old Army surplus diesel generator that I had no idea how to start was hidden in a horse trailer above camp for use in emergencies. Two Porta Potties sat on the east edge of camp, past the last in the line of trailers, tied into the earth with ropes across the top that were anchored to fence posts pounded deep into the ground. A stick leaned against one of the Porta Potties that was used to stir the septic slop to liquefy things. It allowed us to make the interval between cleanings last a little longer; it was expensive to pay a day's salary and mileage for the

Porta Potty man to come all the way up from Lewistown to do the ten-minute job of pumping out the two booths.

. . .

Following the breakup with J and the failure of our captive breeding attempt, I welcomed the work in the wild, where ferrets provide for themselves and I had no role in their life or death. Here, I only documented whether they survived, how they behaved, how many females made it a week, a month, or a season following release, and whether they were likely to produce young to supplement the wild population next year.

Randy gave me instructions on how to turn on the heater in the trailer, the combination to the padlock on the gas tank up the hill, and how to connect the solar-powered battery to the pump to fill my truck's gas tank. He provided me a heap of equipment I would need every night that I stashed in my truck: spotlight, flashlight, data sheets, clipboard, pencils, GPS, pink pin flags for marking where I saw ferrets, Sharpie markers for writing on the flags, microchip tag reader with fresh batteries for placing at burrows to determine the identity of the ferrets.

A few weeks earlier, Randy had finished the late summer work of catching the wild-born young kits of UL Bend, trapping them when they were still with their mothers and marking them just before they dispersed. Randy found another decent crop of about a dozen kits, but that was average. The bigger question was why so few survived to spring each year. Why did the population never grow? Something happened to them over the eight months until spring surveys came around and Randy returned. Was it predators? Was it all at once or did they slowly blink out over the entire winter? Did they move away from where they were born to far-off prairie dog colonies and never return? I was left out here alone to watch them through the fall and winter to try to figure it out.

The ferret search began at dusk each night, a solitary time of brilliant pinks and oranges that faded far to the west to a hollow darkness that was amplified by a flat landscape without edges. The late September air was cool and crisp as I climbed into the cab. Truck keys were always left in the ignition, and binoculars sat at the ready on the bench seat beside me. I turned the engine over and checked to make sure the spotlight turned on and wiring hadn't worked loose. I left the structure of camp and started out in a clockwise motion around the edge of the colony, beaming the spotlight in long arcs in front of the truck with my free hand.

For the next two hours, I rode on a high of excitement in anticipation of seeing a wild ferret. I was hopeful that I could find any of the kits that we released from our failed captive breeding attempt in Malta. It would be a dream come true if I found a straggler from releases in past years that had avoided detection. But tonight the prairie was quiet, nothing moved. After two passes around the prairie dog colonies, I turn on the radio, searching for any station that would come in and take my mind off the fading anticipation that was being replaced with concern that few ferrets were likely to be here. I needed another source of reality to counteract the reality that now, only a few weeks after Randy's surveys, there probably were only four to eight ferrets left on UL Bend. The ferrets from Malta that we released just weeks ago here and to the north might already be coyote scat or owl pellets. I found a Canadian news broadcast from Regina in Saskatchewan: according to a lead Iraqi specialist, HP supplied high-tech computers needed for nuclear proliferation. Date rape drugs were being made of lye and floor tile remover, street name: G and Liquid X. An American farmer in the Midwest was having trouble with high school boys who were smashing mailboxes on Friday nights when they got drunk and drove around in a pickup with a baseball bat. The farmer came up with the solution of buying two replacement mailboxes, a large and a small one, the large one filled with concrete and the small one sunk into the middle of the concrete. The next Friday, a boy swung at the new target and the metal bat bounced back, fatally hitting another drunk high school student standing behind him. The farmer was arrested and awaiting trial.

At 2:40 A.M. I spotted a ferret, emerald green eyes glowing just above the prairie. I threw the truck in neutral and locked my spotlight on it by tightening the clamp on the roof of the truck. I grabbed my binoculars from the bench seat beside me and confirmed by the shape of its head that it was a ferret, likely a small female, peeking out of a burrow mound.

I had seen dozens of ferrets in captivity, cared for their needs, picked up their excrement. I had held their soft, butter-colored slack bodies while I put them under with gas anesthesia, feeling the lightness of a long, thin torso filled with vertebrae and ribs and organs. And I had taken note of their health before prodding them with needles. But this was different. Here they were in control. This ferret told me when I could see her.

She came above ground for an instant, body stretched across the burrow entrance. I thought of Annie Dillard's first encounter with a long-tailed weasel in the Virginia woods, a distant cousin of the ferret that she described as "ten inches long, thin as a curve, a muscled ribbon, brown as fruitwood, soft-furred, alert. His face was fierce, small and pointed as a lizard's; he would have made a good arrowhead. He had two black eyes I did not see, any more than you see a window." This

FIGURE 10. Female black-footed ferret emerging from its burrow. A microchip reader in a nearby box records the ferret's identity when the microchip implanted under the skin above her shoulders passes through the circular antenna and a signal is detected.

female black-footed ferret was even more glorious. She was larger, thicker, a creature of the underground that comes up as a distinctly marked, sleek predator. Rather than the dark brown of a squint-eyed mole, she was a buff lighter than the soil, with hard black lines of a mask, black-tipped appendages. The coloration continues to confuse and amaze biologists as the stripes of a zebra once did. Perhaps the markings are similarly used to confuse predators. Or perhaps they confuse the prey, hiding the dark tender paws, tail, and face from sharp, retaliatory nips of prairie dog incisors.

Noticing me and my disturbance to a clear night, she moved out of view and down into her burrow, away from my persistent spotlight beam, grumble of truck engine, and passing of internal combustion engine fumes. I made a mental note that the burrow entrance was circular and grey, elevated four inches, three hundred feet south of the truck. I grabbed my gear and walked out to the ferret, reciting that the burrow I was looking for was three hundred feet out and four inches high, in the center of the spotlight beam. Walking out on foot, perspectives were altered. All burrows began to look the same on the flat ground where the spotlight beam was splayed out on an area the size of a basketball court with one hundred or more burrows in it. I found what I thought was the correct burrow, shone my flashlight down it, and saw the buff skin and black mask of ferret peering up from twelve inches below ground. I placed my pin flag, obtained and recorded GPS coordinates, and left the microchip tag reader at the burrow. The reader had a black plastic ring that fitted over the burrow entrance and served as a closed-loop antenna. I played out the four-foot wire that attached the ring antenna to an Army ammo box filled with an expensive reader and a twelve-volt battery. Leaving the reader turned on with the ring antenna pressed flush against the burrow entrance, I turned back toward the truck.

I began my walk back to the truck and only got twenty feet away before I heard a beep from the ammo box behind me signaling a confirmed reading. I looked back to the burrow and the ferret was sticking its head above ground, allowing the reader to confirm its identity based on the transmitter chip in its neck. It was a resident adult female known as 388-F, one of four wild ferrets found on UL Bend in the spring, and she was obviously used to the light, the ring, and having her chip monitored. A three-year-old ferret and fourth-generation UL Bend resident, thin and at her lowest weight of the year from the stress of litter rearing, she was a survivor.

CHAPTER 6

Winter

By October, winter had come hard and fast. It froze the water remaining in the bottom of the Army surplus water trailer, meaning the end of running water for the season. I could no longer fill up the hundred-gallon trailer from an old cattle well a mile up the road, drive back to camp, and park uphill to gravity-feed into the camper trailer's tanks. Though it was too alkaline for drinking, the water was fine for washing. Yet from then on, showers would be much less frequent, done by the cupful with water stored in plastic jugs in the warmth of the trailer and heated on the gas stove in small soup pots. I winterized the fragile plastic plumbing of the camper trailer by draining the lines and pouring in a few cups of antifreeze. All washing until spring now had to be done by tub, and waste water thrown out the door rather than down the drain.

Pulses of freezing and thawing mixed with rains left the dirt roads passable one day and choked with mud and ruts the next. The clay prairie soil turned to gumbo, sticking to tires and packing inside wheel wells, bogging down truck frames to the point of making them unturnable sleds of mud and steel, often high-centered in the middle of a road. It would turn an overnight visit into a weeklong stay following a good rain storm. A few weeks earlier, during a two-day rain, biologists from the Smithsonian Institution in Washington, D.C., planned to use a helicopter to visit UL Bend because we warned it would take two days of dry weather to harden the roads. Baffled by the absence of helicopters to rent within a two-hundred-mile radius, they were forced to wait sixty miles away where the pavement ended.

The winter sun rises farther to the south, days shorten, and the opening of elk season brings a peak of traffic to UL Bend. From weeks of solitude, the local population jumps to as many as one or two trucks a day crossing in front of camp, all trying to head south to the tip of UL Bend in hope of a private hunting ground at the most remote corner of the refuge, where there are rumors of large bulls.

The day that elk season opened, just before sunset, a spike bull ran into camp, panting. Forced up by hunters along the wooded river breaks below, he was a fine, tall, young male with spindly, two-foot antlers—surely not the direct target of a hunter. There were no arrows poking from his rump or blood dribbling down his side, yet he was dazed and winded. Rather than flee through camp to higher open plains beyond, he stopped in the middle of the prairie dog colony a hundred feet from the trailers. I told him he was safe here. That I was alone and no one would shoot at him inside the camp, that there was no risk. That he should take a break, catch his breath.

The elk were not always here. Within sixty years of Lewis and Clark's first voyage along the wild Missouri, the development of new, easy transport by steamboat farther and farther upriver brought hoards of well-armed settlers. By 1859, steamboats reached as far upriver as Fort Benton, west of the boundary of the present-day wildlife refuge, and by the 1860s elk were extirpated from the Missouri River breaks of central Montana. In another twenty years, the large herds of bison roaming the upper plains would be extirpated as well. The land was broken up into Indian reservations and homesteads. Central Montana entered into an era dominated by the profitable practice of grazing sheep and cattle and keeping in check wild predators and wild competitors for precious grasses.

Elk remained absent from the Missouri Breaks for ninety years, until 1951, when thirty-one animals were transplanted from Yellowstone National Park. With public lands identified and protection in place, the population steadily grew to 150–200 individuals by 1961 and to the present abundance where they are managed at a density of two and a half elk per square mile. It is one of the healthiest elk herds in the state.

Each year during the early fall, one of the most remarkable yet little-known wildlife spectacles in the state occurs on the very site where elk were first released back in the 1950s. A singular event in the cool mornings and evenings of late September, when hundreds of elk pack into the protected area off limits to hunters along the Missouri River.

FIGURE 11. Male elk bugling during the fall rut at Slippery End, Charles M. Russell National Wildlife Refuge, Montana.

Knowledge of space and place is passed down from mother to offspring that within these invisible boundaries hunters cannot follow. The place is a sanctuary of abandon where females gather in open meadows haloed by golden cottonwood trees still holding their leaves for a last few days into autumn. And then, in the twilight, giant bulls emerge into meadows in display posture, stretching their necks out parallel with the ground to let out a deep guttural bugle.

Up and down the valley, each meadow is a battleground for access to a harem of females. Dueling males with massive antlers like tree branches smell the air, snort, and finally spar with each other for dominance. Smooth wet black noses release puffs of steam. Heavy antlers crackle, tine locked with tine. Thick necks rigid with muscle not only move forward against the force of another male but twist to the side, back and forth, trying to dislodge or unbalance the other. Urine drips from underbellies that show the pulse of hormones flowing through the body that put every muscle and gland on edge. Females with their nearly independent calves watch. Yearling bulls observe how it is done.

Outside of the viewing area, elk are open game to hunters with licenses. Archery season opens before rifle, and those who get there first get the best chance to pick off an as yet unwary prey. The archers travel to the far corners of the refuge on foot to find only the largest bulls with majestic racks that spread five feet wide. They follow the primitive call that those who work the hardest and go the farthest will be rewarded with the least seen, biggest, and as yet untapped. Those who are successful display their conquest with sawed-off antler racks poking out of the beds of pickup trucks for months as they drive up and down the one main street in downtown Malta. In showing off the male dominance of the largest herbivore that still resides on the plains of central Montana, they also show that they hiked in first, got the big kill, hauled it out, and have meat in the freezer. A seemingly more skilled breed than the rifle hunter, archers have to stalk closer to the animal, use strength of arm to pull back the string of the compound bow, with steady hand to hold and aim for the perfect position under recurved stress. Perhaps the females in town will reward him, thinking he would be the most reasonable mate to take home from the bar that night.

Away from town, down at UL Bend, to me hunters were just well-armed migrants. Here for a week with large camper trailers and wives in tow, cases of beer and electric generators. Pestering trespassers when they walked into camp wanting a tow after they had tried going too far off road or made the trip down on the gumbo roads after a storm. They always first tried to bribe me with offers of cash, and when I refused, with cans of beer.

. . .

The harsh, cold wind of late November drove away all but the most resolute hunters, and I made trips outside my trailer only out of necessity. Prairie dogs also spent less time above ground, having long since harvested stalks of dried grasses that they would carry by the mouthful in large bundles back to their burrows for lining den chambers for winter. They do not hibernate, but they can slow their metabolism and stay below ground for days or weeks on end if weather is bad, as it frequently was during early winter in central Montana.

Ferrets similarly decreased their time above ground. While they still must feed regularly, females no longer had kits to raise and, like the males, stayed below ground for three to five days in a row, sometimes only briefly coming up to peer around before hunkering down again for a few more days.

Since the kits dispersed in September, they had steadily been on the decline. Where they went remained a mystery. I saw them for a few consecutive nights and then they were gone for a few nights like the adult ferrets. But for many newly independent juveniles, days turned into weeks, and I could only assume they had died. The high losses, while discomforting, were not uncommon. During the intensive studies of the last wild ferret population in Meeteetse, it was normal for 59–77 percent of young kits not to make it to their first birthdays. What likely happened to them during their quest for a home territory of their own is fairly well known. At Meeteetse (prior to disease outbreaks that eventually extirpated the population), predation was the leading cause of death, with coyotes accounting for at least 57 percent of known mortalities.

Prior to the first reintroduction of kits to the wild in 1991, ferret biologists expected that predation on naïve, captive-reared animals would be high. To counteract this naïveté, researchers built elaborate preconditioning facilities in Pueblo, Colorado. They covered the floor of an old Army building with several feet of dirt and introduced prairie dogs to excavate burrow systems. Then they brainstormed a series of preconditioning tests for young kits prior to release into the wild that included stuffed great horned owls on pendulum strings, badger skin tied onto a remotely controlled model car, and a dog to resemble a coyote. All of this was in an effort to train young kits in captivity to avoid predators from all of the directions from which they might be attacked in the wild: from the air by owls, eagles, and hawks, on the open prairie by coyotes and below ground by badgers.

Dean Biggins and others tested how well ferret preconditioning worked by putting radio collars on the first batch of captive-reared ferret kits released at Shirley Basin, Wyoming. By carefully tracking the movements of ferrets post-release, he found that ferrets exposed to preconditioning were four times more likely to survive than those released straight from the cages in which they were raised. In addition to making the practice of preconditioning a mandatory part of the ferret reintroduction program, Dean's findings showed that 95 percent of the deaths were attributed to predation. Preconditioned animals spent more time below ground and were more able to avoid predation.

In response to the demonstrated threat of predation, for a handful of years beginning in the late 1990s, Randy and biologists at a few other ferret reintroduction sites put up miles of antipredator fence around ferret release sites. They encircled entire prairie dog colonies that were hundreds of acres in size with Australian electric sheep fencing, a tight checkerboard of white plastic strands with interwoven copper wire electrified to keep anything larger than a few inches in diameter outside or inside of the polygon. Any coyotes inside the perimeter were hazed out or shot, and many sites expanded this predator-free zone by shooting any coyote or badger on sight even outside of these temporary plastic barriers. This was done with the knowledge that predators could not be kept away forever; it was only to give the naïve juveniles a head start. Let them adjust and hopefully learn escape burrows before they encounter their first predator, for it takes only a handful of individual missteps or instances of misplaced curiosity to end years of effort to create a small population of ferrets.

Unfortunately, despite the efforts, juvenile survival on sites that used predator fencing didn't increase. Further, where lethal control of coyotes was practiced, kit survival was lower. This counterintuitive finding was difficult to explain. Further clouding the result was the knowledge that coyotes typically don't eat the ferrets they kill; they leave the carcass or just nibble the tender viscera and leave the remains. This suggested they kill ferrets less for food than in an attempt to reduce competition for prairie dog prey, a phenomenon termed by behavioral ecologists as inter-specific competition.

But why did controlling coyotes not enhance survival? Scientists came up with two potential theories. First, lethal control of coyotes could have exacerbated the problem. Although there is no scientific proof, coyote ecology suggests that if you kill the dominant coyote that has an established territory, you invite in new coyotes that are not so well behaved. Thus if a dominant resident coyote who knows its territory well is killed, other coyotes will begin to explore this newly vacated territory, and in their explorations they are more likely to encounter and kill a ferret if it strays too far from a burrow.

The second theory was much more believable and at the same time depressing. Assuming the predator control and fences were successful at reducing predators, we learned that coyotes and other predators weren't entirely to blame for such low success in growing our reintroduced ferret populations. It was becoming obvious that something else was at play.

. . .

December nights lengthened and lingered with the cold and decrease in ferret activity. I drove in circles on the prairie in the cool night without the shine of a ferret's eyes. The silence of crunching frozen prairie at night was no longer broken by the flush of an awakened meadowlark or the gentle cooing of a burrowing owl disrupted from its burrow. The only event started after midnight, when the northern lights appeared, showing waves of green and yellow over the northern hills of the valley.

Pulses of light made me shut off the truck and give in to the otherwise black, clear night. I lingered until the cab was too cold and then restarted my nightly circuits to keep from going numb.

With so few ferrets, each painstakingly secretive at this time of year, I decided to scale back spotlight surveys to the early morning hours. To fill out a full government workday I took on a new project helping Brian Holmes, a graduate student at the University of Montana, who was study flea ecology on prairie dog colonies. Fleas had drawn the interest of Randy and other ferret biologists because of plague, an introduced disease that made it from Asia to the port of San Francisco more than one hundred years ago and that has slowly and steadily been spreading east.

The first image the word *plague* should conjure up is the Black Death that killed one-third of Europe in the fourteenth century, one of three human pandemics that give plague the lofty status as the most catastrophic example of a zoonotic spillover by which all other diseases are measured. Zoonotic because it, like more than 61 percent of other human diseases, originated in wildlife and spilled over into the human race.

Remarkably, despite the prominent role plague has played in human society, we still know relatively little about the disease in its natural environment. The oldest records come from out of the desert regions of Kazakhstan and central Asia (the closest natural proxy to the Great Plains of North America), where plague is widely thought to have originated somewhere between 5 and 38 million years ago. Here scientists have been monitoring populations of the great gerbil for more than sixty years, providing long-term data that illustrate that when gerbil populations booms in wet years, plague outbreaks are likely to follow shortly thereafter, driving the population down precipitously.

To understand this disease requires knowledge of the disease's primary vector: fleas. You first have to accept that fleas are persistently present on many small mammals like the great gerbils of Kazakhstan, only varying their abundance based on a set of factors. One primary factor is the density of rodents on which to colonize and move among—the more hosts the better. The other main factor is weather conditions favorable to survival and reproduction, which provide the ideal humidity and temperature to lay eggs and transition between life stages in rodent burrow systems. Thus warm, wet years favor the growth of gerbil populations on the central Asian steppe and fulfill the two critical needs of fleas: high host density and a favorable climate. In this way gerbils, fleas, and the plague bacteria interact to produce a pattern of gerbil abundance that when plotted on a graph shows a saw-blade pattern of booms and busts. Looking at this graph, one can begin to envision how plague could serve as a sort of "natural" population control for gerbils, the bacteria causing large-scale, epizootic die-offs only when gerbil populations reach some large, unsustainable level. But this idealized vision ignores the biological drive of the bacterium, where like any living thing it shares the common need of avoiding extinction by fulfilling its life cycle of surviving and spreading. Thus, to avoid extinction over evolutionary timescales plague needs never to kill off all of its hosts. To do so would be disastrous. Rather, epizootic die-offs of small mammals appear to be a runaway phenomenon that occurs when host and flea abundance are at their extremes.

Considering the exotic disease, ecologists were justifiably worried for North American fauna when the disease made the recent leap across the Pacific because living in New World isolation meant they had not had time to evolve resistance to the disease like some Old World mammals. In the western United States, biologists quickly found that when prairie dogs encounter the disease it is nearly completely fatal. Spreading cryptically within a span of days or weeks, it leaves managers at a loss as they visit a seemingly healthy prairie dog population one day and find the colony vacant a few days later. Because of the severe threat to prairie dogs, and the role of prairie dogs as drivers of biodiversity on the Great Plains, a group of concerned ecologists headed by Colorado State University ecologist Mike Antolin presented a report at the 2002 North American Wildlife Resource Conference documenting the state of prairie dog decline and the looming threat plague posed. The examples they compiled from across the western United States showed a landscape-scale problem and led them unequivocally to conclude that "understanding the dynamics of plague may be crucial not only for survival of prairie dogs but also for maintaining biodiversity and functioning grassland ecosystems."

To maintain prairie dogs and their functional role on grassland ecosystems, a quick learning curve was required. Researchers stepped in and began to discover pieces of information that could provide clues to how the disease was behaving in North America. They found that the close colonial societies of prairie dogs and their flea communities were ideal for the transmission of plague. Once plague had made it to a prairie dog colony, the resulting epizootic die-off wiped the already small and fragmented populations of prairie dogs (due to decades of poisoning campaigns) off the map. Some populations had small pockets of survivors that persisted not because of immunity but seemingly because they were passed over by chance, providing a glimmer of hope that the handful of remaining prairie dogs would provide seed stock for the slow regrowth of the population. Yet such hope was short-lived as most survivors either dispersed to try to find neighboring colonies or tried to reform the colony only to be knocked back down by another plague-induced die-off at five- to ten-year intervals. These were unstable cycles, because the comparatively slow reproductive rate of the prairie dog was unlike the saw-blade-like consistency of prolific great gerbils in Kazakhstan, and instead declined stepwise in a downward direction toward extirpation.

Looking more closely at these affected areas, researchers found that though a prairie dog colony exposed to an epizootic plague outbreak underwent a 90–100 percent die-off, other mammals, like grasshopper mice, seemed to have some resistance. Further, within the flea communities themselves, some species of flea were better at transmitting the disease than others. Yet for every small piece of information we learned, there was much more we still did not know. How did all of these pieces of information fit together to tell us how the disease persisted and spread on the landscape? Where did it hide between epizootic outbreaks? If we could understand its ecology we might be able to develop theories on where and when it will strike next, or how to control it.

Much of the drive for plague research came from concerns about black-footed ferret reintroduction. Just north of UL Bend, following the release of 167 ferrets over the span of three years (1997–2000) on the Fort Belknap Indian Reservation, an epizootic plague outbreak wiped out almost all of the prairie dogs on the reintroduction areas. It reduced years of work and planning to a quiet prairie of empty burrows where ferrets also succumbed to the disease or starved from lack of food. Some of them perhaps ventured off into the night in search of other prairie dog colonies, only to find miles upon miles of tall hay and wheat fields and grasslands without prairie dogs. Without burrows for cover, they would have been picked off easily by coyotes or birds of prey.

Ten years later, Dave Hanson, a quiet geneticist from University of South Dakota, came out with a plumber's snake and a few pieces of flannel to collect fleas from burrows. He shoved the flannel pieces down the burrows and pulled them out to collect the fleas that had latched on. Taking the fleas back to the lab, he ground them to a pulp and used genetic techniques to evaluate whether plague was still present on Fort Belknap. Dave found that the bacteria were there, at some low background level, and his findings revolutionized our thinking on plague in Montana. Not only did the disease pass through a colony of prairie dogs, causing a near-complete population collapse, but it persisted on colonies through time. His findings suggested that the bacteria might be hiding in the fleas and biding their time until some mysterious cue made them break out again in epizootic proportions.

Just south of Fort Belknap, on Charles M. Russell National Wildlife Refuge, Brian Holmes had similar designs of trying to locate where plague persisted. He came out with his assistants in the summer months to set up live traps of various sizes on the remaining prairie dog colonies, trying to collect fleas straight from their hosts. With the knowledge of which animal they came from, he could go back in the lab, identify the fleas to species, test the fleas for plague-causing bacteria, and if all went well, link plague to certain fleas and their hosts, potentially providing the link between where plague resides and how it moves around on the refuge.

Yet during the winter while Brian was back at the university, no one was there to assess what happened to the flea communities. Randy and Brian wondered what happened to the fleas during the cold winter months. Did fleas become very rare or superabundant on the few warm bodies? Were different flea species present, and if so, did these winter fleas contain the bacteria that potentially were spreading south into the UL Bend ferret reintroduction area? So each week, I placed 150 small aluminum traps with treadle doors in parallel rows on a select group of prairie dog colonies surrounding UL Bend. I opened the small mouse-sized traps at dusk and checked them in the morning after a night of spotlighting, walking the trap line to check for shut doors. Shut doors told me a small rodent had taken the grain bait and was huddled in a corner trying to conserve body heat. Little heat engines with rapid heartbeats fed oxygen to keep the metabolism running at 118 degrees above the outside temperature. I closed the traps for the day so that no mouse or vole strayed into a trap and risked running out of grain bait to fuel its body and freezing to death, and came back out just before dark to reset the trap lines in the evening.

I was told to look for jumping mice and meadow voles, but caught only deer mice. This common, widespread mouse came in clumps, where fifty traps would be empty and then ten traps would be occupied, each containing a single mouse. I worked each filled trap individually, pulling the Tupperware tub of equipment from my backpack, putting a medical mask over my nose and mouth as a precaution to keep from inhaling potentially fatal plague bacteria. I took off my thick leather mittens and then thin inner gloves, putting latex gloves on my bare skin while the rest of me was covered by at least three layers of cotton, wool, and leather.

At ten degrees below zero, it took only seconds for my fingers to become too clumsy to gently hold the mouse and comb its fur for fleas. My numb digits became useless for picking up fleas with tweezers and sealing them in small vials for Brian to identify to species and test for plague later in a Colorado lab. What would take seconds in the summer took minutes. Luckily, the cold weather slowed the fleas down when they were pulled away from the mice and kept them from hopping out of my plastic tub. To keep from losing feeling farther up my arms, I tucked my hands in my pockets to regain finger control, waited a few minutes for them to thaw, and then quickly wrote down the date, time, species of rodent, weight, and number of fleas on Brian's data sheets before covering my hands with my gloves and then mittens before moving on to the next trap.

Driving back to camp, I struggled with issues of scale. The work forced me to wonder whether the small mice scurrying through the burrows on my precious prairie dog colonies had small fleas that contained the microscopic bacteria. It was hard to think of microscopic issues on a landscape that seemed limitless in size. I questioned how such a wide-open and remote landscape could be ruled by such small, invasive bacteria. I thought to blame the bacteria but realized that their effects could be traced back to a fundamental human fault of introducing the disease, and further, that it was not by coincidence that ferret reintroduction sites were in the most remote corners of the Great Plains. These forgotten refugia are about the only places where humans have allowed prairie dogs to persist, and given the decline of ferrets reintroduced farther to the north, and the rarity of ferrets at UL Bend, this is the last great hope for Montana. I thought of how even where we found a pocket of human tolerance, this invasive disease was again bringing the species to the brink of extinction.

. . .

In mid-December, I took a trip to town for the essentials, food and water. Going north from the refuge, I saw fields that showed bald stubble from a summer's growth cut short, the hay dried and put into tall round bales that are unspooled from the back of pickup trucks for hungry northern Montana cattle that have been bred for thick coats and short stocky legs. As I drove north along Highway 191, black Angus appeared as smudge marks on a bald prairie. I came down the hill and into the Milk River Valley where Malta was set; a depressing midwinter prairie town where the browns of fall had faded to grays, and unlike the mountains were not covered in snow. On the few days when snow fell, it was quickly packed into ditches or up against sides of buildings where it had been collected by wind, leaving the prairie bare.

I went into the old office to check my email. It was a weekend, as I had planned, the best time to avoid people. Empty parking lot, building locked and lights off, I found my desk in the corner of the basement just as I left it a few weeks before—again during a weekend visit. I fired up the computer and logged in. There were a few old listserv messages, office memoranda about power outages and potlucks, and then a two-week-old message from J. She invited me to a New Year's Eve party in New York City, a reunion of sorts for friends from the Peace Corps still adjusting to life back in the United States. I thought about replying but instead printed it out. I knew I couldn't deal with the onslaught of questions about us. I knew that seeing her would be bad. Either she would be distant and I would resent her, or she would be kind and I would want her. I knew the temptation of being near her and knew I would try to win her back. Perhaps she wanted me to win her back. But she had rejected me, no longer wanted me, and would never move to Montana. It would be a cycle of hurt.

. . .

By January, when it gets to as low as 30 degrees below zero at UL Bend, it hurts when you step outdoors. With your first breath, the moist air in your lungs freezes to form icicles in your chest. But it is the wind that is brutal, instantly turning exposed skin numb. After a three-day blow, a windless day even at ten below can feel comparatively warm.

On these rare windless days after a snow, there was an opportunity for snow tracking. I would drive out to small isolated prairie dog colonies that are outside of the normal spotlighting area to look for ferret tracks or signs that a ferret had been busy digging out a burrow where prairie dogs rested inside. Leaving camp, the prairie was one large flat white-topped plain forcing me to track the road by memory. Snow crunched under my tires as I drove along in silence, hoping that the snowfall had similarly intrigued the ferrets enough to break their spell of midwinter inactivity and come above ground. Or perhaps the change in barometric pressure from the storm front called on their curiosity and hunger to venture out. Nearing the refuge boundary, my mind searched for any reason to believe a ferret would survive this far out and at the same time leave a small sign of its presence for me in the fresh powder. By February, ferret numbers had dropped to four females and four males that I could find consistently within a week of searching. Finding tracks on one of the peripheral colonies just outside of my normal search area could mean I was missing an animal. The addition of just one new animal would mean more than just one more chance of a litter; it could indicate that there were other ferrets I had been missing, that there was greater reason for hope.

Searching for tracks required careful timing. I needed a full night of little to no wind so that blowing snow did not have time to fill in small ferret tracks or brush prairie dog towns clean of snow from a night of activity. Despite a temperature of 14 below, I enjoyed riding on a four-wheeler out to the corners of UL Bend looking for ferret tracks and diggings. At thirty miles per hour, the wind in my face cut hard, but my core was warmed from the heat that radiated from the small internal combustion engine between by legs. I stopped every few minutes to move my hands and arms and to adjust the cloth over my face. I would turn the engine off and find a silence where I strained to hear something, but there was nothing. Not a bird, not an engine or plane on the horizon. Snow had sealed the earth in white expanses, sparsely spotted with tufts of sagebrush.

I spent all six daylight hours on the four-wheeler, taking advantage of the rare windless conditions and coat of snow, but found no tracks. During the next night, a cold north wind blew the prairie clean, and I was forced to accept that there were likely no other ferrets.

. . .

On January 18, Brendan Moynahan, a doctoral student from Missoula studying sage grouse, drove into camp in his borrowed, brown state pickup. He had flown over the area on the previous day in a rented Cessna to get a general idea of where the birds were and had returned to track his radio-marked sage grouse on the ground. Having finished my morning mouse trap line, I put my cold weather gear back on and climbed into the truck with him.

Brendan had spent the previous two winters at Ferret Camp while studying sage grouse in the area. Prior to his work, no one knew where the declining turkey-sized birds hung out or what they ate during the winter. Brendan was getting the answers and was about to write up his findings as he neared the end of his four-year study. He was visiting only to track the last of his few birds that still had battery strength left in the radio transmitters he had placed around their necks months earlier.

We drove north and east away from the river and off the refuge. Entering the extensive patchwork of federal land near Forchette Bay, we were in the heart of open prairie where Long X Ranch, one of the largest ranches in Montana, used to be based. Founded by entrepreneurial Texas ranchers who aimed to take advantage of the open range of Montana in the early 1900s, the ranch shipped up thousands of head of cattle to Billings and drove them north over the Missouri River. On this as-yet unfenced open range, they settled into a nomadic lifestyle where cowboys pushed cattle over a four-hundred-square-mile area between Glasgow and the Fort Belknap Indian Reservation.

Long X Ranch stories still rattle off the walls of the Hitching Post Cafe on Highway 2 in Malta, where old men sit in corner booths for hours, drinking cup after cup of five-cent coffee. They tell of cowboys on horseback, chuck wagons, outlaws, and drifters. Even former ranch hand, cowboy enthusiast, and pioneer western artist Charles M. Russell was taken with the legend of cowboys at Long X. Russell heard the story of when Long X cowboys were caught unaware by a bear in camp, and without their guns, they had to lasso and stone it to death. He envisioned a scene of rugged beauty where man was pitted against wild nature, with horses rearing and cowboys boldly chucking rocks at the leviathan in his famous painting, "Roping a Grizzly," a then-heroic image that built the mythos of cowboys and sold prints, but now seems brutal and prophetic. In this empty section of the Great Plains, the entire landscape seems domesticated by the extirpation of large predators and their prey.

By 1910, the Long X Ranch was hemmed in by homesteaders who claimed, fenced, and tried to farm parcels of the open range. They forced the ranch to death by strangulation; by 1912, the last free-ranging cattle wearing the Long X brand were captured by homesteaders. Land that the homesteaders settled and then lost over time in multiple droughts is now largely owned by the federal government under the jurisdiction of the Bureau of Land Management. Millions of acres are fenced into parcels that are leased out cheaply to cattle ranchers farther north toward Malta.

We stopped in the middle of Sun Prairie Road, not trusting the snow-packed ditches on the sides. Brendan pulled out the antenna and receiver and stepped away from the metal and electrical interference of the still-running truck. He had learned not to trust a vehicle to start again if stopped this far from civilization and deep into winter, potentially leaving us stranded miles from the nearest homestead with no traffic expected. The signal was a faint pulse over static, still farther to the north, so he detached the antenna and climbed back in.

Opening the door for the brief half-minute to step back inside the cab caused the truck to go cold and fogged up the windows.

He turned the defroster to full blast. "Damn cold today."

He went on. "Probably not as cold as you had it last week, wind was so strong we couldn't fly all week. Wind chill for you at camp must have been 30 below."

"Felt like it."

"Did I tell you of the night it went 30 below at Ferret Camp and I woke up to explosions? One after the other, like gun shots in the trailer."

"No."

"I got out of bed, turned on the lights, but couldn't figure out what was going on. Then I heard another one go off under the kitchen table. I looked down and the soda cans were exploding. Freezing on the bottom of the trailer because the heat rose to the ceiling but the trailer floor was frozen. It was toasty in my top bunk bed at 50 degrees but it was 20 degrees by the floor."

I smiled and gave a satisfactory *humph* sound. My conversational skills were rusty from months of isolation, mind no longer connecting to diaphragm, vocal chords, lips, mouth, words.

"Pray that it does not go much colder than that during the night, because propane begins to liquefy from a gas to a gel. As a gel, your heat won't work." I later found out that for propane to liquefy, it required a devastatingly low 44 below zero—not unheard of in this area. I tried to run through in my mind what I would do if that happened, trying to rig the heater tubing so that I could bring a propane tank indoors. The concept seemed like a worse idea than bringing in firewood and starting a bonfire on the kitchen floor.

. . .

After Brendan's visit, the isolation of Ferret Camp was broken only for a couple of weeks in late February when Randy arrived, with airplane pilot Shaun Bayless, for elk and mule deer surveys. For Randy, it was a much-looked-forward-to chance to spend months out of the office, with his truck, pilot, and laptop computer. One of the few chances during winter when he could get away from meetings, phone calls, and a refuge manager pestering him for annual reports, these annual surveys allowed him to be a wildlife biologist, and return to the roots of being in the wild. Or at least to the best he can be, because the essential field gear of the naturalist had shifted from binoculars, notepad, and bedroll to a GPS unit, digital camera, cell phone, and laptop computer for entering data at the end of the day.

During the daylight hours, Randy and Shaun made systematic transects across the prairie, back and forth at low elevation with strained eyes in a motion that would quickly make me vomit. They looked for a group of brown spots on a brown prairie mottled with white patches from a thin layer of drifting snow. In the evenings back at camp, Randy tied his laptop into the camp phone line that he had rewired through his trailer window to read his email. He replied to people at his own pace, with a whiskey and water by his side. Glancing at the National Weather Service webpage, he checked on any incoming fronts that might ground them from their surveys.

With ferrets and prairie dogs often nestled below ground from the cold, and trap lines slowing down, it was nice to have the company. The presence of Randy and Shaun was a reminder that camp represented the only habitable place to land in the entire area from which they could easily monitor this wide middle swath of the refuge. A reminder that it took a plane to cover all the draws and buttes that expand across this remote portion of the Missouri Breaks.

Their visit also showed me that the prairie dog town had another use, as a landing strip for their small single-engine plane. They landed east to west and then taxied in along the road, carefully crossing over the metal cattle guard between fence posts, and into camp where they tied down the wings for the night in case of a blast of prairie wind. At the eastern edge of the prairie dog colony, I noticed that Shaun and Randy had taken out a one-thousand-foot section of barbed wire fence to lengthen the runway. The fence was strong, tight, but useless because cattle no longer ranged onto this part of the refuge. A group called the American Prairie Foundation had purchased the ranch ten miles to the north on the edge of the refuge. When the ranch sold, the federal grazing leases to UL Bend went with it. Given that the organization's goal was to restore native species including bison, it was unlikely that domesticated cattle with roots in Europe and Asia would ever set cloven hooves on this patch of prairie again. I remembered a time four years earlier when cows with their calves moved to the water wells near camp, filling the air with their soft moans and moos. But rather than mourning the loss of the cattle, it felt good to wonder whether, in my lifetime, I would see bison back on this patch of prairie. I savored the thought of large herds of bison stepping over and around prairie dog burrows that contained black-footed ferrets huddled in the safety of cool, dark chambers nine feet below. I imagined that I would someday need to remember to walk and drive a little more cautiously at night in the presence of such large wild animals with hooked horns.

CHAPTER 7

Spring

Spring thaws came slowly, with teases of above-freezing temperatures that cleared the ground of snow and defrosted the first inch of soil but choked off travel far from camp for fear of rains and muddy roads. In a switch in activity patterns from the summer, details on male ferrets began to take over my field notes as they became more active above ground than the females. Throughout the winter, the four females resided within their well-established territories, recovering body fat and muscle lost from the taxing litter-rearing period. By contrast, beginning in late December and continuing on through April, each male traveled widely across the colony. He marked burrows, sagebrush, and any other object that pushed up from the ground with his distinctive scent glands. Rubbing his neck or bottom to leave the message that he was there, and he would be there again to protect his small group of females.

For the four resident males, there was a critical need to begin the task of taking stock of where the females were and what their reproductive state was. Similar to many members of the weasel family, ferrets exhibit a form of polygamy where a single male tries to mate with as many females as possible. With only four females to copulate with, there was much to fight for as there was only one chance during a single month each year to fulfill their life goal of fatherhood. Further, for a species that typically only lives three to five years, they could only expect one or two years of true social dominance when they could expect to fight off the challenges of other, younger and more virile males.

Despite the high amount of pressure to succeed, direct confrontations between males are rare, and if they occur, likely take place below ground, as they have never been witnessed by a human. For such efficient predators, there is good reason to avoid direct confrontation. Preeminent weasel biologist Roger Powell has appropriately termed the ferret's smaller relatives "hair trigger mouse traps with teeth." For the slightly larger ferret, such deadly incisors and vice-grip jaws could easily turn a social dominance battle into a vicious fight to the death. Ferrets are wired to kill. One time a young female escaped within the captive breeding laboratory trailer in the night and found her way into a bin of more than one hundred hamsters. She killed and ate one or two, then killed the remaining ninety-eight or so, leaving the bodies uneaten. The killing spree was done on instinct, probably taking just seconds for such an efficient and deadly predator, evidenced by the simple death stroke of a neat single poke of a canine through the top of each hamster's skull. In the wild, the use of scent marking likely provides a way for males to express their social status and proximity without risking deadly altercations.

The evolution of scent marking also likely allows males to save energy. Ferrets and their weasel kin are notoriously energy inefficient. Their long, thin body shape, ideal for hunting in burrow systems deep below ground that are inaccessible to other mammalian predators, has a large surface-to-area ratio. This not only means that ferrets have a large amount of skin to lose heat from in the typically cold Great Plains environment, but also that they do not have the rotund shape of a bear or badger to acquire fat reserves for combating lean times of low food availability. This forces ferrets to be master exploiters, feeding as often as they can and producing bumper crops of young during good years, and suffering during bad years when they continually search for food just to survive. This physiological toll and evolutionary niche leaves ferrets in a position where they are continually trying to find ways to conserve precious energy. By marking a territory, the male might be able to conserve energy compared to the need to patrol for intruders. The energy saved can then be used to concentrate on producing sperm and keeping tabs on females.

By late March and early April, my notes were filled with potential pairings. I tried to relate my knowledge of where an individual female was to the movements of the males, guessing which male likely timed his visit to be the one that sired a female's litter. I built theories on which males were likely dominant and which males were trying to sneak in from the periphery. But all of my notes were just guesses, for the

FIGURE 12. Male black-footed ferret at a prairie dog burrow entrance about to pass through the antenna of a microchip reader, allowing researchers to determine his identity. The black mark on his neck was placed by researchers using hair dye for easy visual identification of individuals from afar.

actual act of ferret lovemaking takes place in prairie dog burrows deep underground.

Evidence from captive breeding of black-footed ferrets suggests that the act of ferret lovemaking is vigorous. Some might even term it aggressive. With males having no role in litter rearing and such short life spans, female ferrets have no time to evaluate individual males over the course of the year based on their dominance or suitability as fathers. Instead, it is believed that females test males during coitus, resisting their advances to test a male's stamina and virility. The female's resistance in the weeks prior to estrus is particularly intense, testing the male's resolve with active defense to his advances. This lets time go by as males establish dominance over her territory and burrow in anticipation of the key day when she is ready to breed. As estrous nears, the female reduces her fight and is more likely to permit a male to approach her, eventually submitting and allowing the male to bite onto the back of her neck and mount her for up to five hours of intercourse. Such lengthy and energetically taxing copulation in ferrets is likely a result of the need for a single male to deliver the largest amount of sperm possible. The more sperm he can deliver from multiple ejaculations during the period of intercourse, the more assured he is that his seed will reach the eggs hidden far up the female's oviducts, improving his chances that six weeks later the female will give birth to a group of kits containing his DNA.

. . .

April also was a time when sharp-tailed grouse congregated on lekking grounds—predetermined patches of shortgrass prairie where these birds displayed and bred that were used year after year, generation after generation. All one had to know was the location of these special places to see grouse dance in the crisp morning air and croon at one another. I was given these locations on a road map marked up by biologists who similarly passed down the knowledge of this spring event and recorded the number of birds there year after year. A metric about more than the sexual prominence of a certain area, the counts of the rare annual congregation provided insight into how many individuals still remained in the area, how the population was doing.

For two weeks, I drove out from camp to view these leks. Starting before dawn and taking back roads that were not evident on the land and had to be drawn in pencil on my government map, I stopped at the X on the map and could only hope I was in the right spot to view the lek. I would know only if I were within sight on the specific patch of prairie, or accidentally parked in the middle of the display ground, once the sun rose.

As the sun made its slow ascent, I saw an open patch of ground free of sagebrush and tallgrass on a rise just above the truck. It was a perfect spot for a family reunion of chicken-sized sharp-tails to come together, to show who survived the previous summer and winter, to show who made it from chick to adult, and to show through physical prowess whether they had avoided injury and found enough food to maintain peak body condition to impress the waiting females who stood by on the edge, watching the strutting males. My eyes picked up movement in the soft morning light. Two typically cryptically mottled brown sharp-tailed grouse came into focus. Within minutes a half-dozen were out in the open, flashing their purple patches of throat skin and bright orange eyebrows.

Not far from camp, in a few select sagebrush flats, sage grouse came together in an even grander scene of force. Typically concrete grey in color, the males strutted with heads held high and giant balloons on their necks inflated to display a bright yellow that is hidden the other eleven-and-a-half months of the year. Deep guttural noises emerged from their beaks as they seemingly shook the air out of the inflated balloons with a sound of loudly dripping water.

Because sage grouse are specialists on sagebrush for both habitat and a portion of their diet, as the name entails, there would seem to be a conflict of sage grouse with prairie dogs that girdle and remove sagebrush from their colonies when possible. Yet the relationship is not so straightforward. Although prairie dog colonies are not prime nesting grounds for the grouse because of their lack of cover, the openness of prairie dog colonies can make them favorable as grouse lekking grounds. Prairie dogs provide dependably open space for sage grouse to flock to each spring, and it is often these special traditional spots that are at risk and of conservation concern, compounding the need for prairie dog protection on a landscape scale.

There were bigger questions to answer as to why sage grouse had undergone a 45–80 percent decline and how to fix the problem before extinction. In truth, the story of their decline across the Great Plains is a familiar one, beginning with overharvest that persisted through the 1800s. Without seasons to abide by, the harvesting of hundreds if not thousands of birds exceeded the needs of individual pioneer families. Similar to the story of bison, this type of market hunting was widespread across the Great Plains, slowing in Montana only in 1870 when a six-month summer harvest season restriction was put in place. This moderate conservation step was only improved upon thirty years later at the turn of the century by restricting harvest to a five-month fall season and daily bag limit (meaning how many sage grouse you could harvest in a day) of twenty sage grouse. Despite the newly enacted restrictions, the sage grouse remained the most heavily harvested bird in the state even into the early 1900s, when millions of acres of sagebrush habitat were plowed and fragmented. As the grouse was a prime target because of its size and reluctance to fly when approached, harvest persisted at a high level until the State of Montana closed the harvest season altogether in 1938 and then again between 1945 and 1951. Deciding to finally attempt to survey the population across the state, the state and federal government agencies sent biologists out to count male birds that were on leks in spring. Spring after spring, data came in that could be used to help make management decisions. In response, lawmakers slowly opened certain areas and expanded hunting seasons over time as biologists learned where relatively large populations still persisted. By

2003, the state had settled on a comparatively short two-month season and three-bird daily bag limit.

All told, the damage to sage grouse populations in Montana and elsewhere across the Great Plains was substantial: there was a 50 percent reduction in the range of the species, with many of the areas on their range map becoming vacant only since the 1980s. This told conservationists that, rather than observing a historic collapse, they were witnessing an ongoing decline of the species. Perhaps there was potential for some preventative action to save the species from the fate of the prairie dog and bison. As a logical but seemingly always controversial first step, sage grouse supporters petitioning the federal government to list the sage grouse as a threatened species under the Endangered Species Act. A top-down approach to provide legal protection for the species at a national scale, it would force the state's hand.

It was against this controversial backdrop that Brendan Moynahan had signed on to study the sage grouse of northern Montana. His goal was to help improve guidelines for conserving sage grouse by informing managers about what they needed to do to help grow or at least maintain populations—a bottom-up approach to provide a recipe for success to state government and local landowners. To do this, he needed to assess how habitat conditions were linked to the most important attributes of a viable population: the females. He needed to see how adult survival and the number of young a female produced were linked to their habitat. But to look at how sage grouse females normally behave, he needed to find an area where sage grouse still existed in a somewhat pristine, large patch of high-quality habitat. Thus Brendan headed to UL Bend and the large swath of open grasslands extending north up to Malta and the Canadian border. Perhaps the largest patch of native prairie left in the United States, it also represented the northernmost extent of big sagebrush *(Artemisia tridentata wyomingensis)*—the large, arid prairie specialist to which the sage grouse was known to be most tightly linked.

By following his female sage grouse for four years, Brendan found that they were very sensitive to the surrounding environment. He was able to show that sage grouse that laid nests in areas with lots of grass cover (and thus more insects and seeds for food) were more likely to successfully produce young. Also, he found that the extent to which a small brood of chicks survived to adulthood depended on the year in question. In particular, 2003 was a bad year for sage grouse. Not only was it the winter the soda cans exploded in the bottom of Brendan's trailer, but an outbreak of a new, invasive disease spread through and decimated the population. The parallels with prairie dogs and ferrets were too uncanny to ignore—a familiar formula had again emerged on the Great Plains of exploitation and habitat destruction restricting a wildlife species to a last corner of refuge, and then a disease wiping out the last vestige of hope.

Compared to plague, West Nile virus has fairly humble Old World origins. It was first discovered in 1937, during a survey for yellow fever in the West Nile district of northern Uganda. Researchers collected blood from a woman who described feverlike symptoms, later identifying the virus as unique when the patient's blood serum was analyzed and naming it after its location of origin. The virus was isolated and given to laboratory mice, which revealed the potential for deadly lesions to grow on the central nervous system. However, the patient originally sampled, along with two researchers who handled the virus, all exhibited signs of infection in their blood through production of antibodies but did not show signs of being seriously ill. Subsequent testing of the West Nile district revealed that one in five residents contained antibodies, further suggesting that the virus was just another tropical disease that was fairly common locally, but not often fatal. It was a low priority compared to malaria, yellow fever, tuberculosis, and the host of other diseases of concern during the 1930s.

West Nile virus then took a fifteen-year break from the eyes of science until it was identified during several outbreaks in the Mediterranean. In 1951, 123 of 303 residents of a small town outside of Haifa, Israel, were diagnosed, again with no mortalities and only reports of feverlike symptoms. That same year, several larger outbreaks were observed in Egypt, where young children were most commonly infected. Finally deciding to investigate the disease in earnest, scientists sampled animals in the area and found a high prevalence of the virus in horses, in which it often was fatal. With concern for the pathogen growing, researchers narrowed down the list of potential vectors spreading the disease, finally focusing their attention on mosquitoes—in particular, species of the genus *Culex*, the incredibly diverse (more than a thousand species) and most widely distributed genus of mosquito in the world. Yet despite the ready availability of its mosquito vector, the virus only sporadically showed up in France by the early 1960s, and then in small epidemics in Russia, Spain, South Africa, and India through the 1970s and 1980s.

The virus made the leap over the United States via New York City in 1999, then rapidly spread out across North America and reached

California, Canada, and Argentina within five years. Immediately prior to making the leap to the New World, the disease itself had seemed to change. Beginning with an outbreak in Romania in 1996, the disease had suddenly increased its infection and mortality rate, particularly among elderly individuals. During a Tunisian outbreak in 1997, 173 patients were hospitalized and eight deaths occurred. In 1999, a large outbreak in the Volgograd region of Russia resulted in 183 confirmed cases, of which forty resulted in fatalities. Again more than half of the fatalities were among patients sixty years or older. Either the virus was quickly becoming more virulent, or the hosts more susceptible.

Across North America, the hallmark sign of West Nile virus spread were the dead crows left in its wake. The large, generalist bird was first thought to be the only victim, but later surveys have showed a host of different species from hawks to titmice had succumbed to the disease. In total, according to the U.S. Centers for Disease Control, as of 2013 the virus had been found in at least 326 bird species and counting. During Brendan's sage grouse study alone, between July and August 2003, four of his study birds were found dead and confirmed to have succumbed to West Nile virus. This number might seem small, but he monitored only a small portion of the population on one specific area, and this was over a period of just two months. What did this mean to the broader sage grouse population? Was what Brendan was finding only the proverbial tip of the iceberg? Was the virus almost completely fatal to infected individuals just as prairie dogs were to plague? Were some surviving infection?

That same year, in 2003, human cases of West Nile virus in the United States reached an all-time high of more than 9,862 cases, an exponential increase from the sixty-two cases reported just four years earlier. As the U.S. Centers for Disease Control was scrambling to combat the human concerns (264 of the 9,862 infected individuals died in 2003), Brendan's graduate advisor at University of Montana, Dave Naugle, attempted to put together the pieces of what Brendan and other sage grouse researchers from across the country had found. In a 2004 letter to the Ecological Society of America, he, along with a dozen other scientists, outlined how the invasion of West Nile virus in the Great Plains had reduced sage grouse survival by an average of 25 percent. Further, all 112 collected grouse carcasses had failed to show signs of building antibodies to fight the disease prior to dying. This meant that, like prairie dogs with plague, the sage grouse were not developing resistance. The impact would undoubtedly continue, if not increase, in the future, warning managers that no matter how much effort they put into restoration of habitat for sage grouse, unless something changed, this new disease would always hold the population back.

Unlike the ubiquitous fleas that lived in prairie dog burrows, the mosquito vectors of West Nile virus were not always so abundant on the prairie. Dave Naugle and his coauthors pointed out that dams, irrigation ditches, and ponds that had been created for livestock and farming on the Great Plains over the past 150 years had increased the availability of water in which mosquitoes could lay their eggs and pupate. These water sources also were being utilized by female sage grouse and their broods during the typically dry late summer months, further exposing the grouse to infected mosquitoes. While mosquito control could be an answer in an urban setting to limiting mosquitoes and the risk of West Nile virus exposure to humans, Naugle acknowledged that it was not a reasonable solution across the entire range of the sage grouse across the Great Plains. Making matters worse, previously undeveloped public lands were being explored for oil and gas well development—where energy companies drilled wells and were extracting billions of dollars in oil and natural gas, along with millions of gallons of water that was collected in catchment pools as a by-product. There were thousands of new potentially year-round mosquito nurseries on the prairie.

. . .

After the lek surveys of April, May was a month spent back at camp preparing for the upcoming summer season. On a rare, clear stretch of days when the rain stayed away long enough to allow the roads to dry to a passable degree, I made the trip to Malta to clean out any supplies left at the captive breeding center that would be of use at camp: ferret and prairie dog traps, car batteries, four-inch-diameter pipe, fence posts, latex gloves, medical supplies, first aid kit, fencing tools and hammers, nails, and screws.

I went into the old office to check my email and there was a message from J that was three months old, sent just after the New Year's party.

From: jbb_maine@email.com
To: djachowski@email.com
Subject: New Years
Date: Tue, 07 Jan 2003 11:55:56—0500

PC onslaught, you said it. Friggin' gossipers, there's this circle that people get into asking "have you heard from so-and-so" and "do you ever still talk to so-and-so." Fortunately, I don't, I hardly stay in touch,

FIGURE 13. A spring supply run from Malta, Montana, to Ferret Camp.

> and I couldn't contribute. It made me mad that I am just as much a part of that circle, but surprised that we are still considered together. Just seems so shallow, that folks settle for knowing "what she's doing now", not anything more than that, its just the Job, the Grad School, the Marriage, etc. Although I have to admit, I did enjoy getting the latest from Ashley. I guess I was just hurt that people know about the people I should know, like people in my group.
>
> Again, there's that "intimidating" reputation I've got, and at one point, Lynn, Michele, and someone else onslaughted (a word?) me in Clinton's bathroom when I first met them. I was put on the spot and I guess pretty hurt by it. I also apologized, but they didn't buy it.
>
> I don't care, I don't need them, I don't need anyone.
>
> Woe, how did I get so riled up? I guess just hungry, low blood sugar, PMS or something.
>
> Ignore me.
>
> *Love, J*

I thought to reply but instead printed out the email and folded it into my shirt pocket, not yet having the energy or ability to process its meaning. Everything had slowed down. My thoughts, my work; I was anxious to get back to camp where things were familiar. I logged off the computer, turned off the power strip on the floor, and left the phone number of Ferret Camp on a Post-it note on the computer screen should anyone come looking for me.

FIGURE 14a, 14b, 14c. Two black-tailed prairie dog pups wrestle during the early summer at UL Bend National Wildlife Refuge, Montana.

My truck loaded with gear, I drove south back to camp. Entering the northern edge of the refuge as my truck passed over the welded metal cattle guard between fences, a flock of long-billed curlews with curved seven-inch beaks flew up from the grasslands like brown-feathered flutes.

At camp the sun was out and warming the air to a balmy 60 degrees. Outside my trailer window, green grasses and forbs began to rise from the brown prairie. It was late May, perhaps the prettiest time of year on the prairie, and prairie dog pups had reached the age when they first came above ground and ventured away from their burrow nests. They were wearing soft, fuzzy coats and had narrow heads, round bellies, and short, quick tails that already had black tips and twitched to show emotions of play, anxiety, and curiosity. A straight-up tail meant play or aggression, a downward tail indicated submission or fear, and a waggling tail indicated a gentle approach and greeting with other prairie dogs.

The young pups were shy at first, remaining near burrow openings. Masses of up to eight small young watched the new, multidimensional, bright world and spontaneously broke into fits of play that started with a single individual and led to the whole wad of youngsters tumbling down the mound-shaped burrow entrance over one another.

They tasted grass for the first time as their digestive tracts changed chemistry from the quick breakdown of milk to the slow bacterial fermentation of fresh green shoots. Watching the adults who grazed farther away from the burrow, the pups knew not to venture out that far. They learned the boundaries of territories and their vocabulary changed from the soft squeaks of nursing young to a language of calls. They also learned to look to the air when there was the chirp warning for a hawk and to the horizon for the chirp warning for a coyote. Scientists in Arizona had shown that prairie dog language is even more nuanced, with different warning calls that distinguish a person with a gun from a person without a gun, or even individual people based on their height, shape, and color of their clothes.

FIGURE 15. Two black-tailed prairie dog pups exhibit "kissing" behavior, through which scientists believe they are able to smell each other and gain social information such as whether they are related.

Although I had accepted this old failed homestead turned field camp as my home, the prairie dogs treated it as temporary, building burrows under and around our camper trailers with openings that emerged under the shade of trailers and in the barn. They disregarded me as a threat as I walked among their burrows and into my trailer home, only sounding alarm calls if an unrelated prairie dog ventured into a family's territory. The result of trespassing was swift and aggressive, with a series of calls and teeth chattering meant to scare off the intruder, followed by a full-on attack and chase if it still didn't get the message. Far shorter and more directed calls gave loud warnings to tell the entire colony when a raptor flew overhead or a coyote passed through. After months at camp, I was able to distinguish between the two types of calls and even the sharp chirp for a car coming down the road that, for me, was still out of view. Within the boundaries of family groups and coteries, individuals grazed alone and periodically came together to groom one another. At times they approached one another and brought their small, open, buck-toothed mouths together in an apparent kiss that was thought to be used to smell scent glands, which portray each other's identity and social status.

FIGURE 16. A black-tailed prairie dog pup nursing on its mother.

Prairie dog societies are complex. With cuckoldry (females mating with a male from a different coterie), incest, infanticide, and hierarchies of dominance, their societies rival any Shakespearean drama. Dr. John Hoogland has made a life and career of studying prairie dog societies. Each summer for the past four decades, he has traveled from his suburban Maryland home with family and students to the grasslands of South Dakota, Colorado, or Utah to study the behavior of prairie dogs. Over years of study, he has found that males typically dispersed from family groups by the first year of age, but that for females, a family group of intense, lasting social bonds was both a benefit and a curse. The benefit was from being part of a community where others groom fleas and ticks from you, call out predators, help chase out and bury rattlesnakes, and keep intruders at bay. The curse was being part of a dominance hierarchy that would delay when a young prairie dog could contribute to the gene pool. If you were a young female prairie dog who tried to have a litter too young, it was likely the alpha female of your coterie would kill your young. Sometimes hierarchies would keep you in check for years, up to when you were three years old and closer to the end of your life, before you could rear your own young. One's first instinct is probably to say that within such a tight social hierarchy, those with a low social rank should be the first to move on. By contrast, Dr. Hoogland found that prairie dogs who had intact family structures were the least likely to move on, and that the dispersers typically were those who had lost access to their mothers or siblings. So the decision to move to a new area and risk the ever-present threat of predation is based on familiarity, or the lack thereof.

For an individual prairie dog to start a new colony required daring, calculation, and a bit of luck. For those who did move on to establish new colonies, success relied on the new colony attaining a critical mass of individuals and suitable resources. A team of individuals was needed to construct a burrow system in rapid fashion to avoid predation by badgers and coyotes. A critical mass of individuals would also be needed to have enough eyes on the sky to avoid predation by golden eagles, ferruginous hawks, and great horned owls. But the benefit of being a pioneer was instant social dominance, being the de facto alpha male or female, a key social rank for maximizing genetic productivity.

In camp, one enterprising female had made a special niche for herself. Clearly an alpha from age, she hobbled among the trailers and cringed when her young pups came up to suckle. She was the only prairie dog distinguished by name, Gertrude, a name I had given her in place of the one passed down from Randy and Brendan for her unique tameness to people: Trash Mama. She typically did not move out of the way when I walked past, and she had pioneered a burrow system that led directly into and under the barn where the trash cans were stored. She even had made a neat prairie dog–sized hole in the bottom of the plastic can for easy access to our food scraps, forcing me to hang full black plastic trash bags from the rafters.

When I sat eating dinner outside at the picnic table, she recognized the smell of food and approached, waiting at my feet like a well-behaved dog while tilting her head to the side to look up from below, begging for me to throw her a morsel. I halted at feeding her more than just an occasional whiskey-soaked ice cube out of experiment, yet her trash raiding continued, a diet of plastic and food scraps that, based on her skinny frame and unhealthy patchwork of fur, was clearly a mixed blessing. She surely could not last through another winter, but I wondered if her young would learn her behavior. Was this rare instance of extreme domestication an isolated case of ingenuity, or would the behavior be passed down from mother to offspring? Would our presence create a lineage of Trash Mamas as long as biologists continue to live at this outpost?

In Japan and elsewhere, prairie dogs are popular pets. In Denver, traffic circles are populated with small, urban colonies of prairie dogs that have learned to avoid the highways where lethal wheels zoom by at all hours. In zoos, prairie dogs turn fat from human handouts and even beg for Cheetos from visitors. I hated to think of her as one of these city prairie dogs, but I relished Trash Mama's attention and the insights she provided about the intelligence of this species that surrounded me. Without her tame behavior, I would never have been able to get close enough to know that prairie dogs can discern bull snakes from rattlesnakes, harassing yet avoiding the bite of rattlesnakes but boldly nipping at bull snakes from short distances. She just shrugged off strikes from a five-foot-long, nonvenomous bull snake as she directed it out of her home territory and proximity to her young ones. Her boldness seemingly confused the ambush predator that was otherwise easily large enough to squeeze her to death. Perhaps the constrictor needed the upper hand of surprise to avoid risking being punctured by her sharp incisors.

. . .

For all the activity of prairie dogs above ground, the female ferrets remained reclusive through spring, hiding their young litters underground. Since November, I had consistently found only four females, suggesting that most mortality occurred immediately following litter breakup when kits reached adulthood and dispersed. Randy's population was in a desperate state. Not only were there few females to produce litters, but of that annual crop of kits, few had survived the winter to adulthood to allow the population to grow or even be sustained. It was a chronic problem that had limited UL Bend ferrets despite nearly a decade of reintroductions.

The previous year, during the annual state meeting of The Wildlife Society in Lewistown, Randy presented these facts. He showed that after nearly a decade of releasing ferrets and monitoring the UL Bend population, survival of juvenile ferrets through the winter and to the breeding season next spring when they reached adulthood was only 21 percent. That was far less than the 48 percent overwinter survival observed in the Conata Basin of South Dakota, the only ferret reintroduction known to be succeeding at the time. Further, he showed that UL Bend females averaged fewer kits per litter than ferrets at Conata Basin, producing on average only 2.2 kits per female per year compared to 3.2 for Conata Basin females. For a species with what is known by population ecologists as a "fast" life history strategy, meaning that young mature quickly and need to replace older animals given their relatively short three- to five-year life span, this presented a double whammy. Not only were UL Bend females producing smaller litters, the kits they produced weren't making it through the winter to produce young the following year.

The ultimate question Randy placed in front of the leading biologists in the state who filled the room was this: "How do we get to the level seen in a successful reintroduction site? In other words, how do we increase litter size and kit survival?" This was the million-dollar question that haunted Randy and ferret biologists working on at least five other reintroduction sites where ferrets similarly were having trouble taking hold. Sites in South Dakota, Colorado, Utah, Arizona, and even just to the north of UL Bend in Montana all suffered the same dilemma. They could maintain a few ferrets, but never enough to abandon the need to continue augmenting their small populations with captive stock. Never enough to declare reintroduction a success.

CHAPTER 8

Summer

Mathematical predictive models suggest that approximately 763 prairie dogs are needed to provide enough prey for one female ferret for an entire year. This calculation is based on ferret energy requirements that include not only the needs of the female, but the needs of her average annual litter of 3.3 kits, the male that overlaps her territory half of the time (because he likely has another mate on the side), and other prairie dog predators (that is, raptors, badgers, and coyotes) on her patch of ground. To maintain a population of thirty breeding females, the magical minimum number of ferrets that some federal biologists say is needed to maintain a self-sustaining black-footed ferret population, would require nearly 23,000 prairie dogs on hundreds if not thousands of acres of interconnected colonies.

After nine years and releasing 171 captive-born ferret kits on UL Bend, we had only eight individuals to show for it. Of these eight, the four females were a small group of critical breeders that we could only hope were still alive underground, caring for young litters of small, pink, finger-length kits with eyes still sealed shut. What if these models were wrong and our three thousand acres of prairie dogs at UL Bend couldn't support enough adult ferrets to keep a stable population? Perhaps we needed to quit pumping captive-raised ferrets into the area and blaming poor natural breeding and dispersal and reassess how good the habitat really was. Were there enough prairie dogs to support a population of thirty ferrets and declare success?

Yet to increase the number of prairie dogs by a few hundred, if not a few thousand, would not be easy. We could not follow the approach of painstakingly bringing them into captivity to breed, as was done with ferrets. Rather, we raided existing colonies off the refuge and brought prairie dogs into the refuge. We moved, set, baited, and checked hundreds of traps over and over again with the goal of transferring thousands of prairie dogs from little pockets scattered across the central Montana prairie to UL Bend. By increasing the size of existing colonies and starting new colonies, we hoped that young dispersing ferrets would be more likely to find a home and that these new colonies could act as stepping stones for ferrets to move between the three large colonies already present at UL Bend.

After the long winter, I found rhythm and contentment in long summer days and continuous work of building the prairie dog population. Setting traps, moving animals, caring for those animals, and doing it again the next day. My only company was the activity of prairie dogs, and for the past week, two men who slept in a brown pickup truck parked two miles south of camp on a cliff overlooking the Missouri River. They stopped in Ferret Camp the day after I first spotted them, introducing themselves as the support crew for the Lewis and Clark 200th anniversary reenactors, a team of men that one of the support men, Dennis, bragged "had given up two years of their lives to re-create what was the greatest expedition in American history." I was told the team even included the fourth-generation grandson of Clark himself.

The reenactors were already over a month late in arriving at a site on UL Bend opposite the mouth of the Musselshell River, near where the original Corps of Discovery camped on May 19, 1805. I knew why the team started this expedition but wondered what motivated them to keep going upriver. Had the reenactors' hardship yet shifted their minds from glory in reliving history to simple endurance and pride? Further, what motivated those two middle-aged men to sleep out here in the cab of a truck for days in waiting? Duty? Brotherhood? Regardless, it was good to see someone else struggle, and they were the best type of neighbors, seen only from afar.

Theirs was the only traffic since two weeks prior, when two men came down in a shiny Ford truck full of automatic weapons for use on any living thing as target practice. I heard their shots only when they came near the rim of the little valley where the camp sat. They were traveling the main road due south toward the river, shooting as they moved. They thought they were alone, but really they were shooting just over the camp. I drove up to the top of the hill and angled my truck so that it blocked the road. Adrenalin filled in the gaps where stress and fear should have been when confronting well-armed men so far from the nearest witness.

"What are you doing?" I asked as they came to a stop. The words felt angry and awkward; I hadn't spoken face to face to a human in weeks.

"Shooting varmints," the driver said. Both men were wearing matching black polo shirts with insignias that advertised a gun club near Kalispell.

"You know it's illegal to shoot prairie dogs on the refuge."

"Really, when's the season?"

"Never."

They turned back and I walked north on the dirt road to collect and count the dead animals in their wake before it got too dark to see, knowing the coyotes would clean up the carcasses in the night. I picked up nine prairie dogs and one burrowing owl and saw dozens of holes where bloodied animals had dragged themselves down out of my reach. Tiny crimson trails in the pale soil. At dusk, I returned to camp to call in their license plate to the law enforcement ranger and gave her my tally. A week later I heard that the sheriff in Kalispell went to their homes and that they had to pay a fine of a few hundred dollars for each animal. I still had the ten bagged specimens in my small camper freezer, intermixed with frozen peas and chicken breasts, unsure of what to do with the evidence.

. . .

We were reusing the same Australian electric sheep fence that we had erected to keep coyotes and badgers away from ferrets during the late 1990s to protect vulnerable, transplanted prairie dogs, with the goal of giving the prairie dogs a few weeks of refuge during which to dig their deep, elaborate burrow systems in which they could escape from predators. We strung up miles of waist-high plastic fence around patches of prairie. Energy collected from the sun through solar panels was stored in car batteries and released in rhythmic shocks through thin braided wire. Each day I patrolled the boundary, fixing spots in the fence where deer had pushed through in the night. As I walked, I listened for warning calls of prairie dogs, a sign that they had begun to reestablish the family bonds needed to keep careful, altruistic watch for predators.

After checking the fence for the day, I retreated to camp for a few hours of sleep before starting the nightly task of looking for ferrets.

Later in the season, by mid-August, I was able to escape from this split schedule of day and night work and dedicate my efforts completely to spotlighting and watching the young kits disperse. But during early summer, I looked for the pea-sized reflective green eyes of ferrets, preferably females, so I could count their litters. They, not the thousands of prairie dogs moved, were our only true measure of success—the four remaining females in which our hopes for the future resided.

At midnight, I woke up groggily, grabbed a flashlight, and headed outside. The keys were already in the ignition of the worn-out truck. I pulled out of camp and turned on the one-million-candlepower spotlight as I hit the two-track road. I waved the beam back and forth with my left hand as I shifted from first to second gear with my right. My eyes shifted back and forth with the spotlight. I had memorized each turn and badger hole, allowing my eyes to focus only on the end of the beam splayed out on the close-cropped grasses and forbs. I drove up one side of the valley and down the other, making long loops in the night at seven miles per hour.

They were underground, as they usually were. I knew where the four females were hiding but I didn't yet know if they had bred and were nursing small litters below ground. It would be weeks before I could find out. It was up to the ferret mothers to decide when it was time for their kits to emerge above ground. I was again at the whim of ferrets, caring mothers of young that I hoped were nestled safely in their den chambers below—quiet hopes as I drove loops during a quiet starlit night. I told myself that if I could stay awake until three in the morning, it would be clear sailing until dawn, because at three, by some atmospheric miracle, the AM radio static lifted and replays of the San Francisco Giants game bounced up to the center of Montana, coming in clear for a few hours before fading away with the rising sun.

During night work alone, the mind is prone to introspection. The repetitive movements and solitude forced my mind into destructive thoughts that I usually was able to avoid and replace with the physical activities of the day. I questioned why I was there alone, in a land so open yet closed off to the modern world. I questioned what good I was doing, and I found myself reciting a recurring, self-assuring mantra I had developed and settled on some years back, hidden in my subconscious, only to be remembered at times of such desperation:

> I do it for the guilt in what we have done to the land. For sentimentality and because there is meaning and beauty unique to what was here in the past. For the knowledge that the land has given me so much, and that I need to fix

what has been broken. Restore what has been lost. Re-create some of its wildness. Because the deep truths in life can only be found within oneself following long hours in the wilderness.

Dawn came slowly, and I went back to camp and fell into bed with no ferret sightings. I woke at noon and went outside to sit with the rabbits and barn swallows in the shade of the dirt-floored log barn. I glanced along wall edges to make sure the diamondback rattlesnake that lived in the back corner under a pile of old car batteries was not out hunting. To my memory, the night had passed like a dream, and I could not remember a single song on the radio or the score of the ballgame. The only evidence of the night's work was an empty data sheet with start and end times. Days had run together in my mind and were only numbers now, but I remembered that in the afternoon there was to be a meeting about prairie dogs in Malta, the county seat where ranchers came to buy food and supplies and to fire off their opinions at a bar or government meeting. I ate and put on my uniform for a trip to town.

• • •

I took the two-track road north out of the refuge onto the gravel road that begins near Second Creek schoolhouse. The prairie was rapidly changing from green to the dry brown that dominates the plains for nine months of the year. An hour later, my truck tires spun out rocks held in their treads for weeks and settled into a distinctive hum as I pulled onto the pavement of Highway 191 and picked up speed, heading north to the Hi-Line.

Above the First State Bank of Malta, I entered a meeting room filled with paranoia. Yet another environmental group had petitioned the federal government to list the black-tailed prairie dog as an endangered species. The implications went beyond simple shooting closures and led straight to fear that the federal government would use the Endangered Species Act to dictate how ranchers should manage their land. The room erupted with objections and mumbled threats by ranchers of killing all of the prairie dogs on their land to avoid the potential pending restrictions. To make matters worse, the state was blocking any transportation of prairie dogs. So even where ranchers wanted our help to remove prairie dogs rather than poison them, we were no longer allowed to transplant them onto the refuge to bolster our dwindling populations.

After the meeting, I drove back to camp, unsure of who supported our efforts. The ranchers limited prairie dogs through the plow and poison, the environmentalists brought in heavy-handed protections to fire things up just when we were making progress, and the government layered in bureaucracy and politics. Strong prairie winds blew from the west with no trees to uproot, only toppling grasses and colliding with abandoned homesteads, finding cracks between the logs to whistle through. Back at camp, the wind threatened to break doors off their hinges and drove prairie dogs underground. I waited until midnight, when the wind had died down, to spotlight again. After one in the morning, I finished my first loop of the valley and my mind began to wander:

> Why protect something that is so disliked? Do people really care if ferrets make it back?
>
> In a booming human population, can we ever get back to where we were before?
>
> What do we go back to? When Lewis and Clark traveled across the area, Native Americans still dominated the land, and wolves, plains grizzly, and Audubon sheep still roamed? Before Native Americans entered and there still existed a full suite of Pleistocene large mammals, including giant ground sloths and wooly mammoths?
>
> What is the baseline of wild? How long have prairie dogs been on this patch of land?

. . .

The next morning, Randy called on the phone. He heard the resignation in my voice like a bad hangover. We agreed that things weren't going well and were likely only to get worse. Between the realization that we were likely to have another low production year from only four female ferrets, the exotic disease that was spreading from the north and west and wiping out nearly all prairie dogs in its path, strong political opposition to prairie dog protection, and new threats to creating bigger prairie dog populations even on the refuge, we were no longer sure if we could ever make ferrets stick on this patch of land. Years of work potentially down the drain. He told me to take a few days off and head into the mountains to get out of the dust and heat. I reluctantly agreed to leave camp the next day.

I retreated to my trailer for the afternoon to tie some fishing flies for my trip when the two men in the brown truck drove over the cattle guard and into camp. They stopped in front of the barn and as the truck doors opened, half-a-dozen men spilled out. They stood by their truck and looked around camp, finally fixating on the barn made with large old cottonwood logs faded by time to pale white that dated back to a time before Fort Peck Dam was built and stagnated this section of the Missouri. Back when water flowed down the Missouri through braided river channels and Lewis and Clark's original campsite was not ten feet underwater.

I came outside, and the man I remembered as Dennis proudly introduced the expedition team members who were dressed in an odd mix of neoprene and period garb. All were wearing Army surplus combat boots and eight-inch bowie knives tied to their hips with rope. Their skin was burnt and sun-spotted. Clark's great-great-grandson approached me in tan coveralls made of a thin material like a hospital gown. He extended his hand. It was like a large, open crab claw, permanently molded around the shape of an oar, callused and black on the inside. His hands told me there was something admirable in the quest to re-create the past. The hardship and toil they went through, like me, to try to get back and experience a lost period in time. The reenactors' eyes had a look of exhaustion brought on by hard work and solitude that I knew well. We talked awkwardly about the age and width of the barn timbers, how such large trees or even any trees were rare on this stretch of river today.

A pause in the conversation turned to awkward silence that was exacerbated by the increasingly loud protests by prairie dogs around camp that scolded the group of men. I felt like I should give them a certificate for getting this far or at least a can of beer, but at the same time did not want them to break character. Then, unexpectedly, Clark's great-great-grandson barked a quick sentence like a marching order:

"No chance of getting laid around here."

He turned to me and smiled a broad smile, showing the plaque that had grown between his teeth, making it hard to determine where one tooth stopped and the next began. The men all headed back toward their truck, some of them slowly squeezing in through the gap behind the forward-tilted bench seat. One of the group, dressed in an old, tired windbreaker and baseball cap, came over to ask if he could trade his knife for my engraved belt buckle, which had a silver dollar sunk in the middle, a gift from my uncle for my eighth birthday. Without considering his offer I declined.

I was still standing outside the barn as they headed over the cattle guard and out of camp. More than a year of paddling up the Missouri River and this was likely the most remote location they would ever reach. The closest point to where the landscape still at least partially resembled what it was in Lewis and Clark's time, and yet they didn't linger. The sound of the truck engine faded to the northwest and prairie dogs began to reemerge from their burrows. The patience and beauty of their expedition seemed broken to me. I felt betrayed by the knowledge that they were headed to air conditioning, showers, women, and a steak dinner in Zortman, a small mining town seventy dusty miles away at the base of the Little Rocky Mountains.

I looked out on the prairie dogs that were back to grazing and grooming one another, the hidden ferret population we were struggling to re-create, and found comfort in the thought that I lived in a place still remote and wild enough to throw off these men. That my work restored its wildness even further, making it even more uncomfortable for those looking for the familiarity of city life. I called Randy back and told him I had no desire to leave, that there was still too much work to do.

CHAPTER 9

Chihuahua

Walking into the basement of Mike Lockhart's house in Laramie, Wyoming, can give you the wrong first impression. Framed photographs of black-footed ferrets, swift foxes, polar bears, and brown bears line wood-paneled walls telling of his travels on the prairies, time spent as a biologist in Alaska, and role in ferret recovery. But they are intermixed with calendar photos of scantily clad women in orange and blue bikinis, staged photos of cheerleaders smiling at a camera in support of the Denver Broncos football team. I tried to avoid staring at the photos while Mike showed me to the couch where I was to spend the night. Travis, on the other hand, could not resist.

"Love the photos Mike, you pervert."

Mike let out the soft, self-deprecating laugh that got him through so many heated, small prairie town meetings from Texas to Canada with angry ranchers who opposed ferret reintroduction and the associated protection of prairie dogs.

"They are of my daughter, you jerk."

"What?" We both replied in unison.

"Yes."

"Which one?"

"The blonde one in the middle is Sara." He pointed to one smiling and perfectly tanned model in the middle of a photo of the entire team. We squinted at the photo, looking at her straight blonde hair and toned body and seeing no resemblance to the grey-bearded Mike. As I looked closer, I thought that perhaps a gene did portray the same nose, and another gave her similar eyes. I looked over at Travis at the same moment he looked at me and we were struck by the same realization: she was far too pretty to be the daughter of a field biologist.

Finally Travis chimed in, "Your ex-wife must be gorgeous."

Mike's devotion to ferret recovery was evident in the mileage on his small Toyota pickup truck. That was because he spent most days of the year on the road, traveling with his two Labrador retrievers, meeting with potential and current reintroduction site biologists, managers, and landowners as the black-footed ferret recovery coordinator for the U.S. Fish and Wildlife Service. His task was simple but devastatingly difficult: to downlist and then delist the species from the Endangered Species Act. The latter goal could be reached only when he successfully established ten separate self-sustaining breeding populations of thirty or more adult individuals. His small, unassuming frame hid the power he held in the ferret recovery program, for he was the one who ultimately decided where captive-bred ferrets were released, which sites were started, which sites were kept going, and which sites were let go.

In January 2003, at the annual ferret conservation subcommittee meeting in Fort Collins, Mike asked me to join a core group of ferret biologists on an expedition to look for ferrets in Mexico. My résumé was sparse compared to those of the others Mike invited, but one benefit of my life at UL Bend was that over the past couple of years, I had likely spent more time on the prairie spotlighting for ferrets than anyone else. No other site had a full-time spotlighter to watch ferrets all through the year. It also didn't hurt that Randy planted the idea in Mike's ear, sensing my growing frustration with progress at UL Bend and allowing me to drive my weathered work truck down to Laramie, and then on to Mexico.

By late March, we met at Mike's house and slept in his basement for the night prior to a trip down to the high, arid grasslands of Chihuahua. To ferret biologists, Chihuahua was a place of mythical proportions, allegedly containing the largest prairie dog population in the world, a population of black-tailed prairie dogs that occupied an area of more than ninety thousand acres. At that size, it was almost larger than all existing ferret reintroduction sites combined. The selected ferret reintroduction site, El Cuervo prairie dog colony, was more than thirty-seven thousand acres in size, easily the largest single colony known to exist. Finally, I would see habitat on a scale that offered hope for the recovery program, where ferrets could flourish in their native habitat without intervention.

The existence of the El Cuervo colony was first documented in 1987, and Mike and Mexican biologists began releasing captive-born ferrets there in 2001. Elsewhere, ferret reintroductions had begun to slow. After releasing ferrets for ten years at ten reintroduction sites in the western United States, only one or two of the reintroductions succeeded at maintaining at least thirty adult ferrets. They were both on the largest prairie dog populations known to exist in the United States, in the Shirley Basin of Wyoming and Conata Basin of South Dakota. More troubling, almost all of the large prairie dog populations in the United States had already been used for ferret reintroductions. There simply weren't many large colonies left on which to try reintroductions. In 2001, other than El Cuervo, the only ferret reintroduction site that was initiated that year was just 1,112 acres in size. Following 2001, no new reintroduction sites were initiated until 2004. Worse, eight reintroduction sites that were initiated between 2004 and 2011 (bringing the total number of sites up to nineteen) were, on average, less than seven thousand acres, and most were well below three thousand acres.

Thus El Cuervo meant a great deal to the ferret recovery program. Mike previously had given high priority to El Cuervo by releasing 160 captive-reared ferrets there over the past two years. Surveys by local biologists in 2002 found only nine ferrets. We were heading south from Laramie to try to increase that number, hoping that we could bring some of our techniques and experience when we teamed up with Mexican biologists to help confirm that many more individuals had established themselves, to verify that El Cuervo was the great ferret reintroduction success story we hoped it would be.

. . .

I started out from UL Bend the day before with my government pickup truck loaded down with field gear, including a four-wheeler, ferret microchip readers individually cushioned inside Army ammo boxes, and two dozen four-foot-long wire ferret traps tied on top. Travis Livieri drove down from South Dakota to help mark animals with microchip tags. He filled a minivan with oxygen bottles and anesthesia equipment, ferret traps, towels, blankets, microchips for implanting into ferrets, and Lady Clairol black hair dye to mark the chests of ferrets when they were captured and processed.

Waking early in the morning, we drove separate trucks in a caravan south from Laramie. We picked up Dan, a former eagle trapper friend of Mike's, at the Rocky Mountain Arsenal National Wildlife Refuge just outside of Denver. From Denver we shot east and then south off of the interstate and through the open, mixed-grass prairies of western Kansas and the panhandle of Texas. We tried not to stop as we sped convoy-style through the bleak scablands of Amarillo and Midland, finally spending the afternoon in Deming, New Mexico, where we stocked up with food from a small grocery store in preparation for crossing the border. The store was packed with sunburned greyhairs from up north who were gathering canned goods and pharmaceuticals to take to their hundred-thousand-dollar mobile homes parked in a gleaming mass of aluminum just outside of town. These senior citizens formed their own makeshift retirement village as they avoided the winters of the Midwest like migrant songbirds, moving back north to their real suburban homes only when summer came.

The next day we drove south of Deming through open grasslands choked with patches of mesquite. This corridor down to the border used to be traveled by the last free-ranging bison herd. The herd regularly crossed from New Mexico down into Mexico to the Janos area where we were headed, 120 miles to the south. They had been an international herd that ignored boundaries until 1998, when the Hurt Ranch in New Mexico fenced them into their private land on the U.S. side of the border. With comparative ease, we crossed the border at the quiet town of Puerto Palomas. It was a far cry from the chaos of El Paso and Ciudad Juarez, where ferrets had to wait for hours the previous fall while paperwork was filed and stamped to bring the endangered species across the border.

From the border, we pushed farther south to the small city of Janos, filled up on gas, and then drove the last ten dusty miles farther south to the research station in the village of Buenos Aires. With less than two hundred residents, the entire *ejido* (community) of Buenos Aires was the size of a city block, with small, checkerboard lots forming a distinctive square of human habitation when viewed from above. A dirt-brown stucco house with faded doors and torn-out screens was erected in the middle of each little square lot, the entire community bordered by a dirt perimeter road around the edge. On the west boundary of the perfectly square community, where households gave way to open prairie, a large pile of bleached white cattle bones dried in the sun. Driving through the community, goats seemed to outnumber people in the bare yards surrounded by fences made of tires, wire, and old appliances that were, no doubt, used up elsewhere and brought in just for their retired enameled surface areas, because the modern convenience of electricity had only arrived here two years earlier. The formerly open, treeless skies were now filled with crudely erected power lines strung like tangled kite strings, an effort to connect the *ejido* to the grid of civilization.

The research station was on the north edge of the *ejido,* sticking out from the planned square symmetry of the village like a parasite, further isolated by an eight-foot-tall chain-link fence. After unloading our food and sleeping bags at the research station, we drove out of the fenced compound and Buenos Aires, hitting the main dirt road and heading due south. Our vehicles created large plumes of red dust that rose into the cool, clear March midday desert sky as we headed toward the valley holding the El Cuervo prairie dog colony. Along the road south, in contrast to the tightly packed homes of the *ejido,* we saw sprawling open desert grasslands with orderly farms set at long intervals. The farms were owned by German Mennonites who came down from Canada when they began to believe that their cultural identity was threatened by provincial laws created in the 1920s. Laws that required their children to attend public schools sent out a fearful message of conformity that led the Mennonites to disperse representatives to foreign countries to negotiate a new move. They were looking for a new place to practice their agrarian lifestyle that was removed from modern society and that still had cheap land and few laws of social conformity. After years of searching through Central and South America, they reached agreement with the president of Mexico, Alvaro Obregon, to move their settlements from Manitoba and Saskatchewan to northern Mexico.

The Mennonites brought with them traditional western agriculture, establishing farms on open rangeland that more closely resemble those in Russia or Canada. Planting oats, beans, and corn in the high-plains desert of Chihuahua could hardly be as comforting as it was in the wet, seasonal grasslands of Canada. They plowed the land, found water where they could, and adjusted the timing of planting and harvest to make the best use of brief seasonal rains. Though they used motorized vehicles for the planting and harvesting of crops, many still used horse and buggy for transportation. Women worked in the home and men in the fields, wearing their traditional denim overalls, speaking to each other in the Plautdietsh dialect from old Germany while also learning Spanish and English for the sake of commerce.

They seemed detached from the land, yet tied to it. Many now accepted the new electric power to employ center pivot irrigation systems for their crops. Others moved south to the Yucatán in social protest to avoid the modern convenience, a perpetual cycle of fleeing from the expanding global economy that chases them from one remote corner of the globe to the next. Yet even those who accepted change and remained in Chihuahua were having trouble coping with the growing pressures of society around them. The local news was filled with stories of the Mexican drug war and border drug trade. Armed gangs patrolled the Chihuahua grasslands, sometimes robbing poorly armed and isolated families on their homesteads. Some families traveled south in search of more peaceful lands or to join existing colonies in Belize, Paraguay, Argentina, and Bolivia. Other families stayed put, with some even rumored to have taken on their own role in the drug trade.

Passing the last Mennonite farm, we entered the valley that was bordered by the Sierra Madre Occidental Mountains to the south and east that run all the way to the U.S. border. The area is home to the northernmost population of the small jaguarundi. Its larger cousin, the jaguar, still roams these mountains, occasionally passing through the valley or along the Sierra Madre ridges en route to the border, where it scares Arizona residents with short cryptic visits every few years, the only evidence of its transgression into the United States typically being grainy images recorded on motion-triggered trail cameras.

As we drove along the road, we passed banner-tailed kangaroo rat burrows the size of car hoods, Swiss-cheesed with a multitude of openings in all directions. While spotlighting, we would see the kangaroo rats hopping at such rapid speed that they seemed to hover in the air, their five-inch-long bodies a blur as they streaked by. Only their eight-inch tails were visible, black and white stripes ending in white tufts of fur, as they rushed from under the creosote bushes and mesquite shrubs toward their burrows. Ideal desert specialist, they were storing seeds and grasses for the winter, so that even when water was unavailable to drink they could use the food as fuel to actively metabolize water—creating water from sugar within their bodies.

We crossed over a crude cattle guard; a four-foot-deep open pit between fence lines made cattle turn back, but had two tire-width metal grates spaced at what I hoped was a standard axle width apart. I crept over hoping that I lined up the truck tires correctly, looking down into the cement axle-eating drop below. Once we were over the cattle guard and onto the communal grazing lands the landscape changed from eye-high chaparral and mesquite thickets to open range with sparsely distributed prairie dog burrows telling us we had arrived.

The valley bottom was badly overgrazed and in the midst of a five-year drought that left bare, red dirt to dry in the sun. We saw an emaciated group of cows near a well in the distance. They seemed to be forgotten, left to range freely in the open valley. More often, we passed carcasses of dead cattle that never made it to market, their skeletons still covered with dried skin, their horn sheaths cracked and falling from the bone. The ancestors of the cattle had first arrived in northwestern Chihuahua in 1598 when Spanish conquistador Juan de Oñate led seven thousand head of cattle from Chihuahua City to Santa Fe. These once-vibrant high plains grasslands have, through hundreds of years of intensive grazing, become degraded pasture lands. They were now managed communally by ranchers of the El Cuervo *ejido* who still grazed their cattle here on bare dirt, and after whom biologists have named this, the largest of all prairie dog colonies.

I was struck by the irony of how we name prairie dog colonies after the homesteads and landowners who commonly battle against their very existence—names that represent a perpetual linkage of prairie dogs to human agricultural roots and that persist long after the homestead has collapsed. In Montana, few persons now living can remember the owners and settlers of the Locke or Hawley or Long X ranches on UL Bend, or the locations where these homesteads once were. But the prairie dog colonies that bear their names live on in the stories of biologists who recite them again and again. And because of their lasting importance, the prairie dog colony names on a map will persist long after the log frames and iron plows of homesteads are broken down by the elements and reclaimed by the ground.

We were all in a bit of shock when Rurik List, our Mexican biologist counterpart, met us on the El Cuervo colony with his crew. We were used to open grasslands and close-cropped vegetation of prairie dog towns, but not this level of openness and barren ground. It was a seemingly lunar surface without a root in the soil to hold down the red dirt in a strong wind. With pride, Rurik explained that prairie dogs have persisted here, in contrast to the United States, because the Mexican government did not give the *campesinos* (ranchers) money to buy poison and kill the *perritos* (prairie dogs). They simply had to learn to live with them. Rural communities learning to live with wildlife bring up images of the noble savage, the idea that less-developed human societies live in balance with nature. A spiritual stereotype perpetuated in children's books and by romantics, it was more often based in simple economics. When wildlife is thought to threaten human livelihood, wildlife seldom wins unless the cost of combating it is just too great.

At one time there were 350 *ejiditarios* who were members of the *ejido* that claimed this common grazing land. Each of them was allowed to graze twenty cows on the El Cuervo. But the human population here was declining rapidly, particularly the portion of it that relied on cattle, because as the grass had gone away, the chance of turning a profit had gone with it. At most, two hundred cows were now run in the valley on the communal lands of El Cuervo. The plan Rurik and others had devised was to find money to buy out those remaining cattle ranchers with an annual or lump-sum payment to each and every *ejiditario* and to remove cattle from El Cuervo permanently. It was an ambitious, comprehensive plan that would be doomed to failure in most of the western United States because of the pride of a few stubborn ranchers and government subsidies, but Rurik insisted it would work on El Cuervo.

He explained that the promise of free income for no work seemed like a win-win for the *ejido* of Casa de Janos that manages the pasture encompassing most of the El Cuervo prairie dog colony. Many of the cowboys already had changed jobs or moved on to the United States. Those who hadn't would shift to another form of agriculture on their *campos* or put their cattle in pastures in another *ejido* without *perritos*. Once Rurik and others removed the cattle from El Cuervo, they could let the grasslands recover. They hoped that the pronghorn population, now composed of only two or three groups of a handful of individuals, would recover from its close call with extirpation. Perhaps even reintroduce bison.

. . .

At dusk we returned to El Cuervo for a night of spotlighting, each of us in a separate truck as we drove out to our individual quadrants, attempting to divide the large valley into manageable bits so that we could each make at least three or four loops around our section during the night. Once we drove off the road and onto the grasslands, the smell of the land was different. The thin dust kicked up by the tires was not mixed with the crunching of grasses or forbs as in Montana, and the burrow openings were spread out widely on the open plain, telling us that the prairie dogs were adapting to the lack of food and that the density of prairie dogs was likely to be low. Many burrows were crumbled in or choked with cobwebs, suggesting to us that large patches of the valley were likely to be completely devoid of prairie dogs and unsuitable for ferrets.

The lack of vegetation helped us deal with the enormous size of the valley by allowing for views of up to half a mile in all directions, with our spotlight beams on the flat, open valley floor. Though cattle overgrazing had done its damage, it was the prairie dogs that facilitated this wide-open plain. Unlike large portions of the desert grasslands of the southwestern United States and northern Mexico, here prairie dogs held the mesquite thickets at bay to the valley edges, keeping the sharp-spined and typically unpalatable brush from encroaching and taking over. This behavior has led ecologists to emphasize the importance of prairie dogs for providing islands of habitat in these areas for increasingly rare desert species. Rurik found that at least twenty-one vertebrate species alone used prairie dog burrows here in El Cuervo. That included seven species of rattlesnakes as well as multiple small mammals, and the increasingly rare kit fox.

After four hours of searching, a quarter-mile away I spotted a flash of green. Through my binoculars I saw the narrow gap between a pair of green eyes peering from a burrow entrance. And as I drove nearer, I saw two more sets of eyes and the outline of three young kit fox pups with large ears splayed out wide in curiosity as they stopped their round of play. With their habit of staring at the spotlight like a ferret, it would take another two nights before I would become attuned to the difference between the eye-shine of a kit fox and that of a ferret. To add to the confusion, the eyes of large wolf spiders had the same green glow and low profile of a ferret's. But by the end of the second night I began to get a feel for the scope of the land, how the arid desert comes alive at night.

In addition to kit foxes, large coyotes roamed the open ground, confident in their role as dominant predator here just as they were throughout the Great Plains for the past one hundred years since wolf extirpation. There are whispered dreams about returning Mexican wolves to the area someday. Just across the border, in the Apache National Forest of Arizona, the U.S. Fish and Wildlife Service was reintroducing Mexican wolves. Between 1998 and 2003, at least seventy-four wolves were released into the area that Aldo Leopold so famously tromped around after graduating from Yale in the early 1900s—the very same patch of ground where as a young forest ranger he helped exterminate wolves and saw the "green fire" extinguish from their eyes. It was a moment that he would one day regret and find motivation from when developing the "land ethic," that singular vision of a conservation ethic presented in the *Sand County Almanac* that is still required reading for all young wildlife biologists at universities across the country and world. Rurik and others hoped that, one day, wolves would return on their own to El Cuervo, drawn to the valley once conservationists brought back large wild herbivores. If the rare apex predator could be restored, it would be a final capstone to healing the land, a sign that restoration was complete.

After sunrise each day, we returned to the research station. The cool nights of March kept temperatures low late into the day, and we took advantage by sleeping through morning into the afternoon. After the long drive to get here, our minds and internal clocks were finally adjusting back to a nocturnal lifestyle. We were a quiet bunch as the enthusiasm for finding a large healthy population of ferrets was mixed with the realization that after two nights of searching, we had no luck in finding even a single ferret. We were weighed down by the reality that that there was an overwhelming amount of territory to cover with our spotlights, and by the possibility that all of the more than a hundred ferrets Mike had translocated here over the past two years were gone.

On the start of the third day, when we woke up and sat down to a dinner cooked by two of Rurik's students, Alejandra and Holanda, Rurik informed us that George W. Bush had decided to invade Iraq.

One of the Mexican crew, Jesus, asked us why.

We had no good response.

I was not surprised, but embarrassed. I suddenly understood the appeal of hiding out in the Chihuahua borderlands, a place where Pancho Villa and more recently the Mennonites escaped the past and the demons that accompany civilized society. Here it would be possible to find the freedom to embrace anonymity and never return to a country that seemed bound up by paranoia and a president obsessed with a father-and-son vendetta against Saddam Hussein.

That night we shifted vehicles and I drove around with Mike in his truck for the evening. We worked the dirt roads of the southeastern end of the colony. Although four vehicles were out in the night, we could not see the lights of any of our fellow spotlighters over such an expansive valley. We took a road toward the Sierra Madre Mountains until the valley gave way to foothills and we left the southern boundary of the El Cuervo prairie dog colony. We turned around, taking the same road back down toward the bottom of the valley. By two in the morning we approached the bottom, where our side road met the main road, and noticed a waiting vehicle that turned uphill toward us. Dim headlights set wide apart told us it was an older, larger vehicle, not one of our group's.

"Who is that?" I asked.

"I don't know," Mike replied.

We heard the diesel engine and rattle of the Humvee before we could make out its shape. It was an American military design in drab green Army paint with the back open for bench seats filled with soldiers.

The soldiers drove toward us and then stopped as we pulled to a stop fifty feet away. Dressed in matching green fatigues, they spilled out and encircled us before we even took our truck out of gear, leveling their semi-automatic rifles at the new white government Ford pickup Mike had brought down from Wyoming. I cringed at first for the truck's sake, and then for our lives.

We wondered what was happening in the outside world. As we branched out to help another country, had war severed the diplomatic bond between countries and caused us to be evicted back to the United States, or was this some post–9/11 partnership with Mexico and the United States to stop terrorists on the Mexico side of the border?

Out of the circle, one of the green-dressed men approached in the dark to the driver's side door. He was young, perhaps eighteen, and looking at the circle of others I saw that they were all likely less than twenty years of age.

"*Hola,*" the boy said, and Mike looked at me to translate.

The soldier continued too rapidly for me to translate, and we tried to reach Rurik on the radio. As Mike called out to Rurik, I dropped words like *biologist* and *perritos de la pradera* and *hurón de pata negras*. I tried to tell him about our profession and assure him that our reason for being here at such a late hour was not to avoid being seen by authorities but to find a small animal they had never seen and likely never knew existed. The boy smiled, leaning on the door frame and looking inside, pausing to admire the GPS unit on the dashboard, the binoculars on the bench seat, and the radio in Mike's hand.

After a few minutes of trying, Mike conceded that he could not reach Rurik. He apologized to the boy in English, and the boy smiled and continued to talk rapidly, but did not move. We didn't move. With words failing, the boy's slack posture leaning on the doorsill next to Mike told us that he was not in a hurry. I thought of the fingers on sensitive triggers of high-powered rifles pointed into the cab of our truck. I thought of how stupid this was, to have survived two years in the Philippines avoiding terrorist abduction while American nurses and missionaries were being shot or beheaded by machete on the island just to the south, only now to be shot by the twitching hand of a pubescent boy at two in the morning with American surplus weaponry likely allocated by the American government for the war on drugs, or war on terror, or whatever the latest catchphrase was for the perennial state of alert at this border zone. Here there was a pedigree of arms and conflict dating back centuries. It was a free-for-all zone for bandits and outlaws at the limits of the reach of capital cities over a thousand miles away in either country, where things typically ended bloody in a storm of bullets, memorialized by bullet holes. I thought of the notorious revolutionary and onetime governor of Chihuahua, Pancho Villa, who was mowed down on a hot July day by a group of seven assassins just south of here. Forty bullet holes had riddled the metal carriage of his 1919 black Dodge roadster, and nine bullets had lodged in his head and chest. His still shiny car was one of the most popular tourist attractions in a historical museum down the road in Chihuahua. It celebrated an event rather than the man, whose body was buried far away in Mexico City.

My Spanish failed me and seconds passed like minutes of awkward silence. All we heard was the rough idling of the old Humvee engine as the diesel fumes began to settle. I saw the feet of the armed boys shifting, and the tips of their rifle barrels began to waver in their thin arms. Without warning, Mike suddenly reached into the back for his briefcase, only to return with a flyer that was put together by the U.S. Fish and Wildlife Service in Spanish for the first ferret release at El Cuervo in 2001. He handed the pamphlet to the lead soldier, pointing to the picture in the middle of Mexican president Vicente Fox shaking hands with George W. Bush. Below the photo there was some slogan of partnership written by a Washington political aide. I didn't know if Mike was crazy or a genius, but I imagined his logic went something like this: These young men seem hungry for war, perhaps they will see President Bush and like our decision to enter into one.

The boy smiled, said, "Presidente Bush, Presidente Fox," pointing to each.

He flipped the flyer over and we gave him time to read. We saw him holding it upside down, and I reached to correct him but then stopped as I realized he couldn't read. I wondered what he was trying to prove, if he was trying to impress us or his comrades. I hoped his delay was a good sign that he was attempting to display protocol, perhaps to show to his men that we were indeed official even though in a way that he couldn't discern. Perhaps he was thinking that no drug trafficker, terrorist, or dissident would be so inept as to only carry such a piece of paper in his defense.

He handed the paper back to Mike, and Mike handed it back as an offering.

"No, please keep it," he said smiling a grey-bearded smile and showing humble eyes that I had seen before defuse even the most surly biologists in Montana and irate landowners in South Dakota. Mike looked at me to translate but the soldier understood the gesture, taking the paper and thanking him. He folded it neatly and put it in his vest pocket as he turned to walk back to the Humvee. The soldiers lowered their weapons and followed in behind him.

We sat motionless as the Humvee began a clumsy three-point turn to head back down toward the main road. Still parked, we looked on in silence with our headlights shining on their backs as they rattled down the bumpy dirt road. A dozen soldiers were packed on two rows of narrow bench seats behind the driver and the leader in the front. Young men in the front row rested the butts of their guns on the floorboards with barrels pointed naively at their buddies in the back seat. I now feared for their lives.

. . .

Later that fall, Mike decided to bring us back for a ramped-up ferret-counting effort. During the weeklong spring survey, we found a total of only seven ferrets. All were captive-reared individuals, remnants of the 160 that had been released. We found no sign of a wild, unmarked kit that would have confirmed the occurrence of natural reproduction, that critical step in creating a self-sustaining population. Travis Livieri was again along for the ride as well as my good friend, Doug Albertson, from South Dakota. Mike also invited Drs. Sam Wisely and Rachel Santymire from the Smithsonian Institution, who were conducting a comprehensive study of wild reintroduced ferret populations to look for physical and reproductive anomalies. They were taking samples and measurements for signs of the detrimental impacts of inbreeding because all current ferrets were descended from only eight individuals, looking for the effects of what conservation geneticists would call an "extreme genetic bottleneck," where the genetic diversity of the species was hypothesized to be reduced dramatically during a past period or "bottleneck" when the population size was extremely small.

So far, Sam and Rachel had found no obvious signs of inbreeding depression. Inbred populations can show physical abnormalities such as extra toes or have more subtle cellular deficiencies that could influence their survival or decrease kit production in females. What Sam and Rachel had seen in their travels to ferret reintroduction sites across the Great Plains suggested that ferrets were morphologically fit. They found that a relatively high proportion of sperm were malformed or had kinked tails in wild males, but kit production numbers were similar to those recorded for historic wild populations in Mellette County and Meeteetse. There was nothing that would explain why ferret populations at many sites were not growing.

For the fall ferret roundup, Rurik and Jesus brought ten students from the university in Mexico City, one thousand miles to the south. The influx of students filled the research station bunks. Mike stayed in a hotel in Nuevo Casas Grandes where the crew from the Smithsonian was staying, and because Sam and Rachel had enlisted Travis and Doug as ferret-trapping technicians, the Smithsonian paid for their rooms. As part of talking Mike into bringing me along for this second trip, I promised to travel cheaply. I told him I would not file for government travel reimbursement and would eat food brought from home while staying at the research station. But with students filling all of the research station bunks, after arriving exhausted from the three-day drive, I decided I to just bring out my sleeping pad and sleep on the floor of the research station. I found a storage room in the back of the building away from the activity of the central living room and kitchen. The room was long, narrow, and quiet with windows that looked out to the north. Not wide enough for a bed or desk, it was used to hold boxes of field gear, small-mammal traps, animal crates, and large plastic jugs used for gasoline to minimize the number of trips to town.

I slept until the heat of the day caused the empty gas cans to expand and fill the room with the smell of gasoline. The vapors followed me to the bathroom as I washed my face. In the kitchen area, the students from Mexico City, who seemed never to sleep, were busy making a small feast of paella, enchiladas, beans, rice, and tortillas. I tried to sneak out the front door but they spotted me and invited me to join them for dinner. With hand on doorknob, I agreed to join them after I got back from town. Once outside I breathed deeply, trying to flush out my lungs. I found the keys still in the truck ignition where I had left them and drove the washboard roads into Janos, past the edge of town where, a few days ago, the military man asked for the bottle of Jim Beam he found when he inspected the vehicle. I bought gasoline for the truck and a small replacement bottle of whiskey.

. . .

After the second night without spotting a ferret, Mike asked me to try something new: to get out on foot or four-wheeler and hit the portions of El Cuervo not easily seen from pickup trucks on two-track dirt roads. Given that he invited me and at least in part paid for my trip down here, we both knew this was not really a question. I handed over my truck keys to Sam and prepared my four-wheeler for the night.

After three hours of steering the four-wheeler with one arm, my muscles started to ache even while hardly ever turning. Just holding a straight line on the cold night sapped my strength. I took a bungee cord from the rear luggage rack and tied the handlebars into position, locking the steering straight ahead. With my thumb also growing tired from the sustained pressure on the throttle lever, I fastened the throttle down with bailing wire to hold it at a constant speed. This allowed me to rest one arm while I used the other to move the hand-held spotlight back and forth, switching arms every few minutes.

The cloudless night sky over flat desert land made the Earth's atmosphere seem nonexistent. There was a deep chill of outer space and stars. With no protective cloud layer to keep the heat of day near the crust of the Earth, warmth faded with the night so that the coldest hours were just before dawn, when my energy was lowest. Heat radiated from the four-wheeler engine between my legs but only warmed my calves. My feet were a stage past numb in my thin leather cowboy boots.

After visiting a half-dozen ferret reintroduction sites, Sam was prepared for slow nights of waiting for us to bring in ferrets. By the second night with no ferrets to measure and assess, she raised the ante by offering a shot of tequila to anyone bringing a ferret back to her lab, a makeshift arrangement of medical equipment and liquid nitrogen containers set up in a parked minivan and on the tailgate of a pickup truck in the center of El Cuervo.

At midnight, Doug called in over the radio that he had trapped a kit fox in a box trap he had brought with him from South Dakota. Although it was not the species we were looking for, he said he would bring it in for Sam to draw blood from and later test for distemper virus. Distemper is an RNA virus that can be carried for some time by dogs and other canids like the swift fox, but it is a silent killer of ferrets that have not been inoculated. Prior to plague, which wiped out the prairie dog population and the last few wild ferrets at Meeteetse in the 1980s, serological studies of coyotes from Meeteetse suggested that an active outbreak of canine distemper occurred during the year when that ferret population crashed. Similarly, distemper was blamed for the collapse and eventual death of the last few ferrets harvested in the 1970s from Mellette County, South Dakota, putting the Patuxent captive breeding efforts at the disadvantage of having only a handful of individuals with which to try to start breeding the species back from the edge of extinction. Thus, despite the recent development of a vaccine that was given to all ferrets released back into the wild, distemper still was an ongoing concern, particularly for any wild-born kits that had not been vaccinated.

Arriving at the lab van, Doug deftly grabbed the fox from the trap by the scruff of its neck, putting the small carnivore into a position of instinctual submission similar to domestic cats and dogs when they are grasped by the nape of the neck and lifted clear of the ground. Sam disinfected a small area on the animal's right front leg under the shine of our flashlights. I noticed the small, gentle paws that smelled of earthy, red clay soil. The large head had proportionally oversized ears for both hearing and heat transfer. The fox's eyes watched carefully as hands moved just out of range of its sharp teeth. Sam inserted a needle parallel to the leg just under the skin and hit the vein, filling one vial with blood before switching tubes and filling another.

. . .

By the next night, we still had not seen a ferret. It had been three nights of searching, and everyone made up theories for why our large group of spotlighters had failed to find a single ferret—the cold weather, the drought forcing prairie dogs and ferrets into early states of torpor, the phase of the moon. Given that we were scheduled to leave in another three days, there was also a sense of desperation. Mike sent me to an area where there seemed to be some active burrows with fresh prairie dog diggings at their entrances. It was on the eastern edge of the El Cuervo colony, on land that was owned not by the *ejido* but by the Jeffers family. At one time more than forty thousand acres in size, the Jeffers ranch was purchased in the 1940s by Jayme Jeffers's grandfather. Jayme and his son Billo still ran cattle, but only on a patch of land a third of its original size. They sold off pieces over time and reduced their cow–calf operation as the drought set in and grass vanished.

On the valley side sloping up from their homestead, a yellow layer of dried grass remained to give the land a sense of structure and health compared to the lands used by the *ejido*. Tall cholo cacti reached their thin, branched arms up eight feet into the sky, causing me to have to look forward and steer the four-wheeler for the first time in days. Otherwise my path was painfully straight, doing transects in the night back and forth across the valley.

In response to the chill of the previous night, I put on almost every piece of clothing I had brought from Montana—four T-shirts, three pairs of socks, jeans, two button-up shirts, insulated coveralls, insulated denim jacket, woolen hunting camp, winter gloves and mittens. The wind still crept through as the humidity rose ahead of a cold front from the Pacific.

After two hours, I stretched my legs by standing on the foot pegs of the four-wheeler as I rode, then tiring of that position, began to sit at full stretch on the ammo box lashed down behind the seat. Mixed in with the smell of oily hot air from the engine and dust rising up with my momentum, I got a faint whiff of whiskey rising up from below. For a moment it confused me; then I remembered the small flask I had stashed in the top of the ammo box containing the ferret microchip reader. I stopped to reseal the lid, but first took a swig. The whiskey was warm, sharp, and cheap. Blood rushed to my gut, lungs, head, and I noticed that the wind, which had been at a constant speed, seemed to have picked up as the temperature dropped. Clouds rolled in with the night and brought in the scent of moisture. Snowflakes began to fall, just a few a first, horizontal with the wind. The warmth of the shot of whiskey wore off and a shiver reminded me of the chilled air. I huddled down on the ground next to the four-wheeler, knees to chest and back against the engine out of the wind. I watched as the flakes began to fall at a faster rate and accumulate around patches of grass, humps of dirt, anything giving them purchase.

To get another blast of warmth, I took another shot of whiskey, but this time the only warmth was that coming off the engine through the back of my denim jacket. My eyes felt heavy and I decided to wait out the squall. I woke five, or maybe thirty minutes later; I couldn't tell but I had grown stiff from the cold. The snow had stopped and the air was still. I wondered what time it was, looked at my watch, but did not remember when I first shut my eyes. It was 3:20 in the morning. Spotlight still in hand, I turned it on without getting up. A thin white layer of snow sat on hillocks of grass, sparkling like crystal in the bright, unnatural light. Out of spotlighting reflex, I scanned to the east and then back to the west. I started to stand and peered again the east, catching the emerald green eye-shine of a ferret not thirty feet away. I crouched back down in astonishment, staring at the little animal as it stared at me, poking its head just outside of a burrow rimmed high with red dirt surrounded in a white backdrop. I quickly set a trap at the burrow and left as fast as possible, lingering only a moment to make sure I had set it perfectly, knowing this might be my only chance for a ferret. I drove away, deciding to stay away for an hour so that I would give the ferret time and space to think of entering my trap. I cruised down to the south and wondered if I should track down Mike to tell him of my find. Thoughts of glory passed through my mind, Mike's admiration and a shot of tequila from Sam. But what if I didn't catch him? What if my find was not the first? I completed a loop of the pasture and checked on my trap after only thirty minutes, stopping fifty feet away, and slowly walking up to it. The cylindrical trap was four feet long and four inches wide, with one end stuck straight into the burrow entrance. The portion that rose three feet out of the ground was wrapped in a protective layer of burlap, just as I had been trained, to mimic the darkness of a burrow. The treadle was two feet from the end and set on a hair trigger, allowing the long body of a ferret to enter the trap entirely before the trap door at the bottom sprung shut.

I took my flashlight and peered down the trap, seeing a small female huddled at the far end. The face was finely marked, the gentle browns counterbalanced with a stark, black mask. I saw her rounded ears and wet black nose, and eyes that peered up at me and then turned away as she slunk her ten-inch body around in the narrow trap. She turned again and chattered at me, the loud, guttural, shrieking chatter that always pricked my nerves and set my senses on edge. I looked up from the trap and to the west, thinking that any other spotlighters within a mile would hear the sound on such a still night. I lashed the closed trap to my four-wheeler and headed toward Sam as the rush of adrenalin mixed with the calming satisfaction that at least one ferret still existed in this valley.

The aggressiveness of the ferret told me she was wild. Captive-born ferrets were raised to be moved by trapping, used to being confined, and typically were quick to resign to being in a trap and transported. Similarly, wild animals that had been trapped before seemed to give up the battle and refrained from surprising the trapper with chatter and a spray from their scent glands. This one seemed naïve to my harmlessness, ready to fight to get out of the cage.

I took the female ferret down to Sam, who had heard the four-wheeler engine approaching and was standing by the pickup truck tailgate. I showed her the filled trap and she let out a hoot and rushed off to start up the minivan engine to heat the cab. Rachel turned on lights and prepared the anesthesia equipment. I handed Sam the ferret and let them go to work. It would take a half-hour before they were done with the animal, so I headed back into the night, hoping that the drop in barometric pressure and moisture of the storm would have brought other ferrets to the surface—hoping that I spotted the animal because of the change in weather rather than cold, whiskey-induced luck. When I completed a loop and returned to the lab van Sam was smiling, and I knew even before she told me that it was indeed an unmarked female, the first wild-born ferret to be found in Mexico. A sign of hope, followed by the sweetest-tasting shot of cheap tequila I would ever know.

. . .

At dawn after the last night in El Cuervo, I loaded my four-wheeler in the back of my truck and strapped it down for the long drive back to the research station and then home to Montana. In the two nights since finding the first ferret we had captured an additional ten ferrets, two of which also were wild-born. Not a bad run, with the final night bringing in five of the new animals, but it was still far from being a robust, healthy population. Ferrets were too few and far between compared to the Conata Basin in South Dakota where reintroductions were succeeding. There, for the amount of time and effort we put in, a good haul would have been three times as many ferrets.

Back at the research station, when I joined the end of the line of vehicles, I immediately noticed that my truck had a new sound to it, a hard rattling of broken metal. I stopped and crawled underneath, tracing the sound to a broken shock absorber behind my left front tire. I thought back to who borrowed my truck during the night while I was out on the four-wheeler. I had given the keys to Mike, who had then passed them on to one of his technicians at the captive breeding center in Wyoming. She had come down the day before with the last batch of captive-reared ferrets for the year, bringing the total released over three years to 239. That night she stayed up to try to see a wild ferret, but she clearly didn't nurse the old Ford over the desert washboard roads.

I limped the truck back to the research station at ten miles an hour, pulled into the research center compound, and got out my socket wrenches. With the wheel off, I found that the shock absorber was fine, but that the metal arm holding it in place was broken. A three-inch wide piece of steel had sheared from with the continuous vibration of traveling for years on dirt roads in Montana, and now Chihuahua. I felt the years on the truck, on my life, during these travels to the ends of the Great Plains, the distance it had taken me away from the woman I loved, and the dependability it showed me in winters far from help. I was tired and knew I could get it home to Montana with only the leaf spring holding the front axle onto the chassis, so I worked the layers of hardened prairie mud off the bolt heads just to remove the dangling parts.

By the time I had the truck jacked up and ready to remove the part, it was no longer morning. The sun was already high in the sky, and I took off all my outer layers of clothes down to a T-shirt as I crouched in the soil under the edge of the truck in the warming air. Behind me a scrawny calico kitten snaked under the chain link boundary fence. It gave a half-cry that changed to a purr as it rubbed against me. I cringed from the feel of the sickly animal on my back and stood up, pushing hard on the socket wrench but unable budge the bolts loose. I found a piece of pipe in the back of the truck to use as extra leverage and grabbed an old can of sardines used for baiting foxes. I opened the can for the kitten as I sat back down in the shade. Putting all of my weight on the pipe, I popped one oversized bolt loose and then another. I paused between bolts to breathe my last day's worth of desert air, wipe the sweat from my eyes, and watch the kitten feast on the oily, brown fish. The single tin can held more food than the kitten had seen or likely would ever see at one time in its short, vagabond life. I wondered why I hated most house cats so much but tolerated this little varmint.

After putting the wheel back on the truck, I went inside and found that everyone was asleep. My gas-filled room was already hot and toxic, so I found a can of tuna and made a sandwich. As I finished, Rurik came into the kitchen.

"You're up early."

"Just fixed my truck. Broke off a metal supporting arm last night."

"Do you need a mechanic?"

"No," I said, reaching for and holding up the broken metal pieces. "I think it will make it back to Montana."

"We can go to a metal shop. The Mennonites can fix anything."

"I don't want to put you out and make you drive me a long way."

"Oh, it isn't far. Just down the road." I had trouble believing him, knowing there were only a handful of Mennonite homesteads within ten miles of "just down the road" and wondering why a local rural homestead struggling to avoid modern society would have electricity and expertise to work on such capitalistic contraptions. My curiosity was piqued.

"If it isn't too much trouble," I said.

"No trouble at all, let me get my keys."

Five miles down the road we pulled into the compound of what looked like a normal Mennonite farm but that Rurik insisted was the local metal shop. He explained that most Mennonite communities have at least one specialist in just about everything: one plumber, one mechanic, one teacher, one nurse. We walked out of the sun into a large metal building and found a flurry of activity. Young men dressed in identical denim overalls pounded and scraped and heated metal at different stations and corners of the building.

An older man was bending a metal bar in a vise on a workbench in the middle of the shop. We walked up and he spoke to Rurik in Spanish. Rurik pointed to the metal pieces in my hands and I held them up for him to see.

"What is that from?" He asked me in perfect English.

"The arm that holds the shock absorber for my truck. It just sheared off last night."

"We can weld it together," he said, and turned to one of his apprentices, shouting orders in the old German Plautdietsch dialect.

Within three minutes it was back in my hands and we were on the road back to the station. I was amazed by the ingenuity of the Mennonite community, their stubbornness to stay on this dry land. The scope of the land as it played out before us was made more impressive by the self-sufficiency of its people. Yet at the same time, I cringed at the thought of them succeeding to the point of buying more land and draining the water table by putting in more center-pivot irrigation systems for their potato fields. They were not immune to the forces of capitalism and the free-market economy, but conservationists were no strangers to these laws as well. We used wealthy donors to buy out ranchers or pay them to keep their cattle off the land, a grand financial scheme of using money to pay people to do nothing and thereby preserve a chunk of land that is unique. There seemed to be some basic economic law that was being broken that would prevent Rurik's biosphere project from operating in perpetuity, but my mind was addled from days on bumpy roads and a lack of sleep.

On the drive back to the research station I asked Rurik, "Why isn't this working?"

He looked over at me, not understanding what I was getting at.

"The largest p.d. colony in the world and so few ferrets . . ."

We both knew the answer was, "I don't know," and we drove in silence to the turn for Buenos Aires.

"Perhaps if the rains come next year, the prairie dogs will come back, and so will the ferrets," he said.

I thought about the bare valley floor and wondered whether there was root stock and a seed bank of grasses left to sprout if and when the rains came. Would they be able to create a biosphere reserve that lasted long enough to remove cattle and wait for the prairie dogs to return? I wondered how many more ferrets Mike would allocate here in the hope that prairie dogs would come back.

"I hope so," I said as we pulled to a stop at the research station.

The following year, only a single ferret was seen.

CHAPTER 10

Conata Basin

It was one o'clock in the morning, but there was still a steady stream of headlights as cars traveled down to Pine Ridge Indian Reservation from Rapid City.

"Heading home from the bars or on liquor runs," Travis said. "Pine Ridge is dry as a bone."

Upon arriving in Conata Basin, adjacent to the seldom-visited South Unit of Badlands National Park, spotlighters are told not to stop. Never pull up next to a stopped car or pick up a late-night wanderer. A veteran ferret wrangler in the area, Travis took the extra precaution of being armed.

I was in the Conata Basin of South Dakota for the fall ferret survey. By invitation only, you provided your own truck, gas, and lodging. You paid for your own meals and found your own hotel accommodations, but the benefit was the promise of seeing more wild ferrets in the span of a week than most ferret biologists get to see in a lifetime. Certainly more than I saw in the years of surveys I assisted with at UL Bend and in the few short weeks I spent in Chihuahua the previous fall.

The large size of the ferret population was mostly a result of the vision of Bill Perry, the U.S. Forest Service district supervisor. Walking into the office Bill ran just off of Main Street in Wall, I instantly knew he was different from other managers. The unmistakable aroma of ferret was pervasive when I stepped in the door to the Forest Service office, emanating from his room in the back where he kept a "pet" black-footed ferret.

In the mechanic's bay, where the district fire truck would normally be parked, sat a cage of prairie dogs, tame enough to pet. Out back in the fenced parking lot there were a half-dozen green Forest Service vehicles, and each truck had a hole drilled through the roof where a spotlight stuck through.

An unassuming dark-haired father of two high school–aged daughters, Bill could easily be confused with a high school principal rather than a leading conservation biologist—someone who took the countercultural stance of unabashedly liking ferrets, and perhaps even more so liking prairie dogs, with the same fervor as would a thoroughbred horse breeder. He used that same heightened level of obsession and agricultural-type animal knowledge to deduce that his delicate critters required a large swath of contiguous habitat rather than a few isolated patches of prairie dogs on the edges of his district. They needed thousands of acres occupied by prairie dog colonies to reach his goal of a ferret Eden. To achieve this required cunning and vision. First, finding an area in his district suitable for prairie dogs where they already existed in large numbers, and second, arranging for land swaps and alteration of management plans to prioritize the prairie dogs' conservation for ferret recovery. His proud creation was known as the Conata Basin, a now almost contiguous block of more than thirty thousand acres of interconnected prairie dog colonies largely on federally managed land on which he could try to restore the ferret.

We arrived in the town of Wall at dusk after eight hours on the road and checked into Anne's Motel. Without stepping foot in the room, we filled up on gas and drove south through Badlands National Park and into the basin as the sun began to set. We met Bill on the side of the road, where he threw a dozen traps and two readers in the back of each of our pickup trucks. He handed us data sheets and maps and told us to get going, saying that we could chat later. He wanted us to catch the first flush of ferret activity right after sunset, when eager young ferrets take advantage of the cover of darkness after being underground all day.

I headed south and west, deeper into the badlands formations and away from the paved road. Arriving at night, I could not get a feel for the land. Flat prairies were fragmented with a maze of vertical badlands formations. My spotlight shone brightly off the white sedimentary walls. Between those steep walls, prairie dog burrows were densely laid out across the flat basin bottoms, easily twice as dense as those at UL Bend. Swift foxes and badgers also were more plentiful, darting around with their similarly green eye-shine. Off the prairie dog colonies, grass grew higher here than at UL Bend, dense two-foot tufts without sagebrush, taking advantage of the increase in rainfall that occurred as one moved farther away from the rain shadow of the Rockies. Even on the prairie dog colony, the carpet of golden mixed-grass prairie hid fissures in the soil we would later name "axle eaters," for their sudden way of bringing a truck to a dead stop in an instant and threatening to tear the axle right from the frame. Without the moon, I could not tell north from south or decipher where I was on the photocopied topographic map Bill handed me that was a dizzy assortment of lines making fine, concentric rings.

I drove to the center of the prairie dog colony away from the badlands. In the middle of the little side basin there were dried-out reservoirs dug into the earth by man decades ago. I dodged an old burned-out car with inch-wide bullet holes ripped through its sides by 50-caliber machine guns. The harmless, rusted piece of metal was used as target practice and now served as a reminder of how at one time this land was overlooked by our government, only of value as a World War II gunnery range for the U.S. Army Air Corps. But locally, rather than forgotten, war over this piece of land continues. Once considered the northern part of Pine Ridge Indian Reservation, it was withdrawn from the reservation by the U.S. military back in the 1940s for the war effort. Now these few rusty pieces of shredded scrap metal sticking out on an open prairie were reminders of how some wounds remain unhealed. This was just one in a long and continuing series of land grabs by the federal government, taking most of the land away from the tribe in a quick instant and then forever after reducing the remainder one section at a time.

A late-night thunderstorm rolled through and we huddled down. Rather than call it a night and head to town, Bill told us to stay in our trucks and wait it out. From long experience, he advised us to stay on our designated parcel of habitat but to be near a paved road in case the rain really got going and things "got sloppy." I turned off the spotlight and truck engine as the raindrops began to fall. They came gently at first, with the pace of a slowly moving storm system. The cab of the truck was warm, and after the long day I began to doze off. The truck was parked facing south, and I began to think of the reservation boundary just a short distance away. It seemed like the boundary of an unfamiliar country. Thoughts of the present were mixed with guilt for the past. By the time of the last major battle between government troops and the Sioux in the Battle of Wounded Knee in 1890, the weapons switched from guns and ammo to politics and treaties, schoolhouses and missionaries. Children were taken away from their parents, hair cut, language curtailed, and religion morphed into Christianity. Just a few miles away, decades of persecution reached a bloody climax with a singular massacre that also marked the end of the Wild West, after which there would forever be a new ecology of the Great Plains: laws, greed, and food rations.

The rain picked up to solid drumming on the metal roof of my truck and I fell deeper into sleep. I dreamed of Stronghold Table just to my west and the armed escort that park biologist Doug Albertson used to say was required if a Park Service employee ever needed to go there. It is an area of Badlands National Park dominated by tables and valleys, far removed from the paved loop road along the northern walls and pinnacles familiar to the park tourists. From the top of the table, there were spectacular open views all the way to the Black Hills, but this place was made inaccessible by an uneasy custody battle between the National Park Service and the Oglala Sioux Tribe. The standoff between the tribe and government continues to persist through time in an otherwise domesticated West. I imagined the armed tribesmen still were there, ready to defend this most spiritual place, where the last ghost dances were performed prior to the massacre of Spotted Elk's band along the banks of Wounded Knee Creek. I fell asleep and dreamed of a few enraged men filtering down from their perch at the table and onto the flatlands after noticing my bright spotlight wandering on the otherwise dark prairie. They approached my truck head-on in the driving rain, only instead of a group with automatic rifles, I saw a single man approaching my truck in the darkness. It was Chief Red Cloud walking straight into the beam of my headlights, asking why I was out here, what I was searching for, why I needed to find something so badly at the cost of making crisscrossing tire tracks on the prairie and filling the night with my light pollution. Was I truly being selfless and doing this for the ferrets, or for myself?

. . .

At 4:00 A.M., I woke up from slumping across the bench seat of my pickup truck. Bill was on the radio. He said the lightning and rain had passed. The roads were fine, but he told us to avoid the gully crossings where water had run off the steep walls of the badlands and flushed through to the south. He reminded us that ferrets "dance" after rain storms, a euphoric jumping behavior as they arch their backs and leap into the air, sometimes chasing their own tails. It was an occasion to let loose the cooped-up, underground energy that was perhaps awakened by the storm. Or perhaps it was a drive to come above ground to explore following the change in humidity and barometric pressure. Even over the radio, Bill's excitement was infectious. His desire to see these little carnivores back on the prairie was shown by his actions and by the expanse of prairie dogs in the basin. In the years to come, ferret populations would grow at a site in Wyoming and one in Arizona, but Conata Basin was the first, and at this time the only, sign of hope.

I started up the pickup, turned on the spotlight, and got back to work. I began to take off cross-country to view my assigned territory diagonally and suddenly caught something in the headlights. I had previously seen only a single ferret all night, but after the rain, I traveled only twenty feet and nearly flattened one. The ferret ran across open ground and I caught it with my spotlight as it moved out from under my truck. A near-$10,000 accident, I said to myself, remembering all the toil captive breeders go through just to raise a single animal.

All of a sudden, ferrets were everywhere. Each time I got back into my truck after setting a trap on one ferret, I had to drive only a quarter-mile before spotting another. A rush of euphoria was brought on by seeing such a rare animal in abundance. It was a window into a past world that was bountiful and complete; wildness that I had been searching for validated by numbers. A measure of what natural looked like. I began to go into a competitive zone like a teenager in an arcade, where I lost awareness of everything else and concentrated on the goal of tallying more ferrets, gathering more data. Bill gave a running tally of his, and others', ferret sightings on the radio and thereby created a spotlighter competition to help him evaluate his progress in rewildling the basin. By the end of the week's friendly competition I would finish second to Norm Eisenbraun, the trusty field hand of Bill's who knew every bump, nook, and cranny of the basin. An honorable loss in what Bill had dubbed the first annual "Ferret Roundup."

On the final night before we all departed back to our separate parts of the world, we drank to celebrate Norm's victory at a bar in the town of Wall. We were finally able to see the other spotlighters' faces after a week of solitary ferret hunting. Randy and I happily joked about the magic of South Dakota. That this was how ferret spotlighting was supposed to be, nights full of activity. There would be uncertainty about where a ferret would pop up rather than the sense you had at smaller sites where you knew where the five or six ferrets were and it was just a matter of when they came up. The Conata Basin ferret population hovered at just over two hundred individuals, larger and more stable than the wild populations monitored during their final years in Meeteetse, Wyoming, and Mellette County, South Dakota. We drank into the night and eventually hit on the question that any ferret biologist eventually asks after visiting Conata Basin for the first time: How did this work? Why was this site successful when others had failed? Why did Bill have so many ferrets here yet there were only five in Montana, only eleven in Mexico, only thirteen in Utah, and only seven in Colorado? Further, annual surveys and litter counts by Bill and Travis showed that the population was still growing, but why?

The obvious answer was size. With prairie dog colonies covering an area of more than thirty thousand acres, the Conata Basin was home to one of the largest prairie dog populations in the world. But how many prairie dogs were needed? How big of an area was needed to just maintain fifty or a hundred ferrets? Based on how often ferrets needed to feed and the density of prairie dogs, predictive models suggested that most of the reintroduction sites should have at least thirty breeding-aged adult ferrets, but almost none did other than Conata. Did territoriality limit ferret populations? But the largest prairie dog colony in the world, El Cuervo, where ferrets had the most space to expand wasn't working. Perhaps the prairie dogs and their burrows were just too few and far between in that desert grassland to make good ferret habitat? Why were litter sizes of ferrets smaller and survival rates for dispersing young lower at UL Bend than in the Conata Basin, even though the Montana site seemed to have vacant habitat?

On the long drive home to Montana, I set my mind to getting at these questions by breaking ferret success at Conata Basin down into its incremental parts. To shed light on our troubles at UL Bend and elsewhere, I would have to get to know a few ferrets well. Study their fine-scale behaviors, litter-rearing practices, how they used and competed for space, how many young they bore, and how close the young stayed to their mother's territory. In essence, how black-footed ferrets were supposed to be.

. . .

The next spring, with a small amount of money and my old Montana field vehicles loaned to me by Randy, I started my research. To maximize the number of ferrets I could watch, I focused on a single, highly productive portion of the Conata Basin that had been one of the original release sites for ferrets a decade earlier. Just a relatively small patch

FIGURE 17. View of the badlands formations that bound the Conata Basin of southwestern South Dakota.

sheltering 2 percent of the larger basin prairie dog population, the North Exclosure was believed to be one of the best corners of ferret habitat in the basin, as evidenced by Bill's records of the area consistently being occupied by females and their young each fall. By the time I arrived in 2005, the site contained wild-born ferrets at least three generations removed from their original captive ancestors.

It was called the North Exclosure because it formed the northernmost portion of the Basin on an isolated bench between two gullies leading up to the boundary of Badlands National Park. Bill told me that because of its isolation, I would have to rent a farmhouse twenty miles away and drive south in my trusty old Ford pickup over the Badlands wall each night to get down into the basin. I would then head due north, off the blacktop and onto the prairie through a barbed wire fence gate along a dirt two-track road. After following the fence line north for a couple of miles, I would need to cross two steep gullies that were impassable following a rain storm before arriving on the North Exclosure prairie dog colony.

I started in June when the females were holding tight underground with their newborn young, The first night out spotlighting I was filled with anticipation like a student at a new school. I was desperate to meet my new ferret friends that I would spend so much time with over the coming months. I knew their lives would be much more interesting than mine as I became their follower, forming attachments to the point that their plight would similarly dictate my mood. A female ferret losing a kit to a coyote would send me into a momentary depression, the digging of a badger at a den burrow entrance would fill me with anxiety. I tested the balance I needed to maintain between the detachment required to observe their natural behavior and the obsession needed to stay awake all night and follow their drama. As a scientist, I also wondered whether there would be enough animals to study. If Bill's projections were correct, the North Exclosure should have contained at least three to five females that would produce litters that fall, the bare minimum needed to discern broad patterns in their behavior. I circled around on these thoughts over and over again as I explored the five-hundred-acre prairie dog colony, learning the shape of the North Exclosure and finding the ideal route by which to maximize my coverage of the area while spotlighting in long loops through the night. Anticipation and concern had built by dawn when the sun finally rose, forcing me to note my stop time and come to the realization that my data sheet was blank. Not a single glimpse of a ferret.

The second night, at 2:57 A.M., I sighted a ferret. I placed a microchip reader at its burrow entrance to see if it had been marked by Bill and Travis with a microchip tag, letting me know where it was seen in the past. I finished a loop of the exclosure before returning to find the reader registering a serial code that verified that ferret as 04–013M. The 04 told me it was first captured and likely born in 2004, the 013 told me that it was the thirteenth ferret captured that year, and the M that it was a male. At 4:30 A.M. that morning I spotted another ferret and got a reading of its microchip: 02–047F. It was a female originally born miles away on the other side of the basin. I breathed easier, knowing at least one male and one female were present.

The third night, I found and confirmed another female, 04–019F, and two more males, 03–034M and 04–006M. Although it was a decent number of ferrets for three days of searching, I worried that two males and three females might not be enough to complete the study. The three males, though interesting, were of less importance to me because a single male in a small population can breed multiple females, making the presence of multiple males seem superfluous. Also, male ferrets have no role in litter rearing. This meant that watching the behavior of males during the summer was uneventful and provided little insight into why this population had such high productivity. Biologically, we knew that females were key to a healthy and growing ferret population, and having only two of them to study all year would make anything I found unique to those two females. And it would be tougher for me to later make that always important scientific leap to say that I found a pattern that could be generalized across the broader population.

I returned home at dawn and thought of switching to a different part of the basin. I pulled out one of Bill's maps to look for possible alternatives, but there were no sites that had such a sure-fire history of containing ferrets year after year and at the same time were isolated by gullies on three sides so I could be fairly certain the ferrets would stay there with their litters year-round. Besides, if I failed to find a better spot and decided to return, during the time away I would have missed some of the behaviors by the two females. I decided to stay put, but it was unsettling to think that as it stood, a year of work, a year of my life, depended on me learning as much as I could from just the two females, focusing my life on the handful of minutes I would stumble onto these two animals when they finally decided to come above ground.

. . .

In addition to trying to document the where, when, and how of ferret activities, I needed to understand their prey if I were to understand *why* ferrets were behaving the way they were. Like their taxonomic cousins the weasel and stoat, ferrets are high-energy carnivores. Laboratory studies have shown that they require nearly continual access to food to keep their fast metabolisms burning, so how that food is distributed should matter, especially if they require larger areas when there is less prey. Also, theory would suggest that if food were scarce, the ferrets would likely be more territorial when prey density is low. If I could find out whether either of these patterns holds true, and at what thresholds they come into play, I might begin to explain the lower-than-expected densities of ferrets at UL Bend and other reintroduction sites across the West.

I already knew that the task of mapping the distribution of prey would be difficult. Even in the relatively small area of the North Exclosure, thousands of prairie dogs moved about and came in and out of burrows. Rather than undertaking the ridiculous task of counting individual prairie dogs, I decided to map their burrow openings. Where there were more prairie dogs there should be more burrows, and I could therefore use the density of burrow openings to gain an approximate picture of where prairie dogs were most abundant. Yet there were still likely more than twenty thousand burrows on the colony, and no easy way to map them other than by visiting every burrow with a highly accurate GPS—a seemingly endless task of roaming back and forth across the prairie dog town to systematically collect longitudinal and latitudinal coordinates for every individual burrow with a precise instrument that communicated with satellites orbiting above, allowing me to record the location of each trashcan-lid-sized burrow opening within an accuracy of a few feet. The end result of four months of work would be a sort of prairie dog road map showing all 27,890 burrows as dots on a computer screen. The darkest clusters of dots would be where I expected prairie dogs to be the most abundant, and thus my ferret mothers to be spending the most time.

Getting that close to at first hundreds, and then thousands, of prairie dog burrow entrances taught me that each burrow was slightly different, telling a different story of what was underneath. Burrow entrances where prairie dogs had built up the outer edge of the entrance like tall, volcanic cones told me that the burrow was built up like a chimney over a vertical tunnel below that shot straight down for perhaps nine feet before veering horizontally and connecting to chambers and other burrow systems. Burrow entrances that were built up into rounded mounds led to similarly complex burrows below ground. These mounds had slanted entrances that were easier for a prairie dog to stand on when surveying for predators upon first coming above ground. Like master engineers, prairie dogs use the interplay of low mound and high chimney burrow openings to facilitate catching and forcing winds below ground—a flow-through air ventilation system. Although usually interconnected below ground, burrows within a couple of feet of each other might not be connected, but I never found a burrow entrance off on its own. This told me that almost all prairie dog burrows had at least one other exit point, even if the other entrance was only an inconspicuous escape burrow that was difficult to find and more than twenty feet away, flush to the ground.

. . .

I split my activities, mapping burrows from dawn to midday, then falling into bed by mid-afternoon, only to slowly wake up and head out for a night of spotlighting a few hours later. During the day, staring down at the prairie for hours a day forced me to speak to the ground. In my head, I rattled off the names of the low grasses and forbs that sounded both descriptive and reminiscent of the human imprint on the land: buffalo grass, western wheatgrass, sand dropseed, prostrate vervain, needle-

and-thread, red three-awn, Japanese brome, tumblegrass. Only two months earlier, the secrets of the grasslands were hidden in drab browns. Then spring rains cued root stocks and young seeds to reveal themselves. Black Elk saw the spring rains as visits from the spirit world when "the grass shows its tender faces." Its growth brought multitudes out of hibernation and drove migrations. All life here is timed to spring rains. If the story of climate change is true, scientists predict that the most endangered birds in North America are not going to be the colorful, migratory songbirds of the East Coast, but the bobolinks, dickcissels, and other grassland specialists of the western prairies.

Spotlighting every night on the same piece of ground, I began to know the pattern of things. By the middle of June I learned that certain ferrets came up to look around at the same hour each night. Some ferrets seemingly came up randomly, and others like clockwork, only once every three days. I also learned that the burrow occupied by a female ferret three days ago is likely to be where she is now. This was largely because in June, kits were still young and had yet to open their eyes and feed on meat. They nursed, minimizing the amount of time a mother was away from her kits. I went for days without seeing the two females, and then they suddenly appeared at their same individual burrow entrances where I had seen them four days earlier. But then, a week later, I saw that the families had moved to new burrows a short distance away.

Unlike most mustelids, ferrets change their den sites frequently. Ferret cousins like the wolverine are notoriously picky den site selectors, finding the perfectly fallen tree or conglomeration of overlaying boulders under which to raise their precious young each year. Or in the arctic, digging long, complex snow tunnels—one of the reasons some argue that the threat of global warming could eventually harm the size and distribution of wolverines across the northern range. Yet the ready availability of so many potential den sites in the form of burrow chambers premade by prairie dogs seemed to drive the ferrets to favor only temporary housing, changing these critical nurseries for their young multiple times within a rearing season. Exactly how frequently they moved their young to new den sites was still relatively unknown. On June 28, 1982, Tim Clark had observed a wild female in Meeteetse transporting three kits over a distance of sixty feet to a new burrow location. The female gripped each still-young kit by the scruff of the neck gently between her sharp canines and hauled it to the new den—the entire process taking three trips and fifteen minutes. Apart from this sighting, the earliest that ferret kits have been seen in the wild above ground was in mid-July when they were more than half-grown. Clark noted that "nothing is known of BFF [black-footed ferret] development between birth and appearance above ground."

. . .

June 18, and a coyote and badger crossed onto the North Exclosure for the night. The short, squat black-and-white badger waddled among burrows with the lanky coyote trailing just behind. I stopped the truck and panned the spotlight to watch the badger as it moved between burrows, smelling the entrances to check for the heat and aroma of a sleeping chamber filled with a prairie dog family before moving on. After visiting eleven burrows, it set to work on the twelfth using its powerful paws and inch-long claws to dig into the opening. With strong, thick forelegs, the badger quickly enlarged the hole, digging the packed dirt and thrusting it rearward under and behind its body directly between its hind legs.

Unlike the narrow-faced European badger, the American badger is built like a stocky torpedo, so once the hole was wide enough for its head, the whole body could follow in quick fashion. Within ten seconds, the entire animal was underground and digging deeper toward the prairie dogs' sleeping chamber. The coyote paced and then sat down at a nearby burrow ten feet away, waiting in case the badger flushed a prairie dog through a connecting burrow and out to the surface.

This partnership was not unique to Conata Basin. Native American societies of the desert Southwest and Pacific Northwest told stories of the friendship between coyotes and badgers. More recently, naturalists and farmers occasionally reported the coyote and badger traveling together as early as the 1920s. The obvious reason for such a pairing was that both species would seem to gain a benefit. Yet to ecologists, the partnership of predators searching for the same prey was atypical. Examples from Africa to America all suggested that two predators going after the same prey should be in aggressive competition. Lions kill hyenas, cheetah cubs, and any other competitors. But here, on the Great Plains, two predators with different techniques had evolved to at least temporarily tolerate each other. They almost fit that anthropomorphic term of being friends, each gaining some benefit from the other. The badger stirred the sleeping prairie dog up toward the surface, and the coyote hazed the prairie dog back underground. A mutually beneficial partnership that was only validated by science in the 1990s, when research showed that coyotes killed 34 percent more ground squirrels when they worked together with a badger than when they hunted alone.

I thought about how such a partnership could decimate a ferret family, causing a mother or her kits to run to the surface into the jaws of a coyote. Not that the coyote needed help; some reports suggest that more than 60 percent of ferret deaths are caused by coyotes. I began to do the math in my head of how many times I saw a ferret at a certain burrow and came back the next day to find the burrow dug out by a badger. There were only a handful, and of those I had seen, I confirmed that the ferret was still alive a few days later. Perhaps the duo had yet to perfect the craft of ferret hunting.

I wondered what partnerships might have been lost when ferrets were extirpated and whether they had been restored. Did ferrets have unique predation strategies in South Dakota that were lost when the Mellette County population disappeared? Did we now have ferrets that are attuned to hunting only in the style of Meeteetse, Wyoming? From the Meeteetse seed stock, we knew it was common for ferrets to avoid the crepuscular (dawn and dusk) times of day when coyotes are most active. The ferret liked to hunt in the middle of the night when it was safest to move from burrow to burrow, like a badger, only without the need to dig. It caught prairie dogs while they were sleeping, biting the neck and bending its long body to press its back against the wall of the burrow, locking the prairie dog in place while its canines held on and the ferret strangled the prairie dog to death. Less commonly, there were daytime hunts, when the ferret crouched by an entrance in the morning, peering down the burrow as the sun rose and a prairie dog began to make its journey up into the light. When the prairie dog came to the lip of the burrow entrance, the ferret would pounce down the burrow, taking the prairie dog head-on by surprise, just below the surface. Perhaps other ways of hunting prairie dogs have been lost.

In all likelihood, the response of ferrets to prairie dogs has been so ingrained over thousands of years of co-evolution that it has been passed on in their genes and quickly reinstituted when captive ferrets are released into the wild. Prairie dogs certainly did not forget the absence of ferrets. I remembered one morning at UL Bend when the warning cries of prairie dogs told me that there was something unique in camp. I looked outside and saw a young male ferret running above ground in the midday sun. It ran between burrows and was quickly mobbed by a group of three adult male prairie dogs and forced into a burrow. When the ferret poked its head out, the prairie dogs made lunges at its nose and then backed away, trying to force the ferret farther below ground so they could plug the burrow with dirt, burying the threat in a tunnel that would become abandoned. It was a behavioral response similar to what I had seen by prairie dogs toward a rattlesnake, when prairie dogs hazed the deadly rattlesnake below ground and blocked the burrow by filling it with dirt—making a seemingly impenetrable four- to eight-foot-long earthen plug. Despite the ferret's recent extirpation from the prairie, and being absent for multiple generations of prairie dogs, it seemed the prairie dogs inherently remembered their black-masked foe.

. . .

On the third of July, I found a new ferret, 02–015F, bringing my total count of adult females on the North Exclosure to three. The shy female had remained hidden from me for a month despite my passing right by her burrow every night. Seeing her made me doubt what I really knew about these animals. Was I missing other ferrets, and if so, how many? At the same time, she reassured me that I was getting closer to understanding how the North Exclosure ferrets were behaving. She was located apart from the other two females on a large northern chunk of the prairie dog colony that seemed seldom used, close to the boundary fence with Badlands National Park. She filled that last vacant piece so that the colony was divided into thirds, with females occupying the lower, middle, and upper parts. The three females appeared to be ideally spaced so that no large patch of habitat was unoccupied, fitting what ecologists refer to as an "ideal free distribution." The theory was that all available habitat should be occupied optimally, so that resources are utilized in proportion to their availability—a theory of symmetry and efficiency, where competition is minimized and fitness is maximized.

Life took on a rhythm of mapping burrows by day and then returning to the colony to spotlight at night. I learned the timing of each of the three females. Female 04–019F, who lived on the lower end of the colony, came above ground almost every morning between 1:30 and 2:30 A.M. Female 02–047F, who lived in the middle of the colony, came up between 3:30 and 5:00 A.M. Female 02–015F, who lived on the upper end of the colony near the park boundary was more secretive, coming above ground and being visible briefly once every two or three days for a few minutes sometime between 1:30 and 5:00 A.M.

On July 15, I saw the four kits of 02–047F for the first time. They were already two-thirds the size of their mom, poking their heads above the ground. Sometimes all four kits poked up at the same time through

FIGURE 18. Normally active only at night, this female black-footed ferret stayed active above ground at dawn near the border of Badlands National Park.

the narrow burrow. The next day, I saw 02–015F move her three small kits by the scruff of their necks just as Clark described in Meeteetse, dropping them one at a time into a new den burrow where she had stashed a freshly killed prairie dog. The new den burrow was in the section where I was to map burrows the next day, and by midday the next morning I reached the burrow where I had left a reflective marker during the night to mark the location of the den. I used the GPS strapped to the four-wheeler to log its longitude and latitude. The ferrets were far underground, out of view. I imagined them sleeping together in a den chamber. The little kits would be fat and happy after eating the prairie dog carcass down to nothing, leaving only the tough skin and skull nearly picked clean, brain removed. Sharp, new, gleaming white teeth were put to use as the kits slowly weaned off of mother's milk and began a full meat diet.

By late July, the prairie had turned dry and the heat from the day lasted late into the night. Long stretches of cloudless days burned unexposed patches of skin, and I quit day work by noon not out of exhaustion, but out of risk of heat stroke. I returned to the colony only after midnight when the air had cooled. While the prairie wilted the kits were growing quickly, becoming increasingly hungry. Mothers needed to hunt and bring their kits to new kills every two to three days, so I saw them above ground more often during the night. Kits now were able to occasionally follow their mothers in moves to new den burrows rather than being hauled, tracing their mothers' above ground movements nose to tail single-file, running across the prairie. When Bob Henderson saw this in the 1960s in Mellette County, he called it "train behavior." Clark similarly observed ferret trains on occasion in Meeteetse. It was more evidence that these ferrets in the Conata Basin were behaving like their wild ancestors.

• • •

On July 25, I checked the reader placed on a burrow entrance at the lower end of the colony where I believed I saw 04–019F the previous night. After nearly two months of monitoring the same half-dozen animals, I had memorized the nine-digit code for her microchip tag, but this time, looking down at the digital readout screen, I was confused. I went back to the truck to cross-reference the number with the list that Bill and Travis had provided me of more than one hundred ferrets they had marked in the basin. I scrolled through the numbers, hoping that it was a new male just passing through the colony, but I knew something was wrong. The ferret I saw the night before was small, surely an adult female or kit. But a kit wouldn't have a microchip yet. I switched to the list of marked females in the basin and scrolling to the bottom of the third page found that the number belonged to female 03–031F.

The appearance of a new ferret after nearly two months of monitoring led me to further question everything. Had I been missing this female? Had she avoided my hourly drive-bys each night? Worse, because she was near 04–019F, had I been confusing the two for weeks? I felt my grasp on this small ferret society slipping. I had thought females established and defended territories, with little tolerance for female intruders. My theory of ideal free distribution, with ferrets democratically claiming habitat and avoiding other ferrets, seemed to be shot.

I called Dean Biggins, a federal ferret biologist with experience dating back to the Meeteetse days. He told me he had seen this type of close spatial overlap between females once before in Wyoming. He went on to tell me that more often ferrets were likely territorial, following the pattern common in mustelid ecology, where females protect hunting areas to ensure the ready availability of prey to meet the energetic needs for raising kits. Females with poorer access to prey must travel greater distances to hunt, be more exposed to predators, and as a result have a harder time raising offspring to adulthood. There seemed to be great incentives for females to protect patches of good habitat and be aggressive toward one another, but the presence of 03–031F and 04–019F on the same night only a few hundred feet apart didn't fit this model of behavior.

Perhaps it was an exception to the rule? By the first of August, 03–031F and 04–019F were still staying in burrows within three hundred feet of one another. They were out hunting and I saw two kits in the area. Whose kits were they? Were they 03–031F's? Why would a mother ferret move between colonies while rearing her litters—a seemingly unwise trek through a deep gully and tall grass ideal for coyote stalking and disorientation, with no burrows to find refuge in? And why would two females tolerate living in such close proximity? Perhaps 03–031F is 04–019F's mother, and she is helping or at least tolerating the presence of her kit from the year before.

This uncertainty haunted me until August 11, when I saw that the two kits were sharing the burrow of 04–019F. By then, with almost all kits spending time above ground in the night, the absence of kits with 03–031F told me that she did not have any or lost them earlier this year. Perhaps she lost the kits as early as mid-July, which would explain her late arrival at the North Exclosure. If a female ferret loses her litter, it would make sense that she would attempt to find a new area with better access to prey or farther from the predator. Moving on seemed like a sound evolutionary strategy.

To identify an open territory or establish a territory of one's own requires luck and perseverance—luck in finding a piece of high-quality, vacant habitat; perseverance in claiming and protecting the spot of land from other females. But at the same time, it appeared that the females relaxed their firm boundaries to others. Based on my stippled burrow map and observations so far, it appeared that space use overlap was most likely to occur where prairie dogs and their burrows were most plentiful, suggesting that ferret behavior seemed to follow the common ecological and socioeconomic theory that when resources are abundant, individuals tend to be less aggressive toward one another, more tolerant of sharing.

But as a general rule, despite these ferret territories slightly overlapping each other in areas of high prairie dog density, I rarely saw two females raising their young in close proximity to one another. When female 02–047F, who occupied the middle of the colony, was in the southern part of her territory, female 04–019F, who occupied a part of the colony below her, usually was not on the adjoining northern portion of her territory. Thus ferrets might have overlapping territories over the course of a season, but perhaps never be in close proximity at

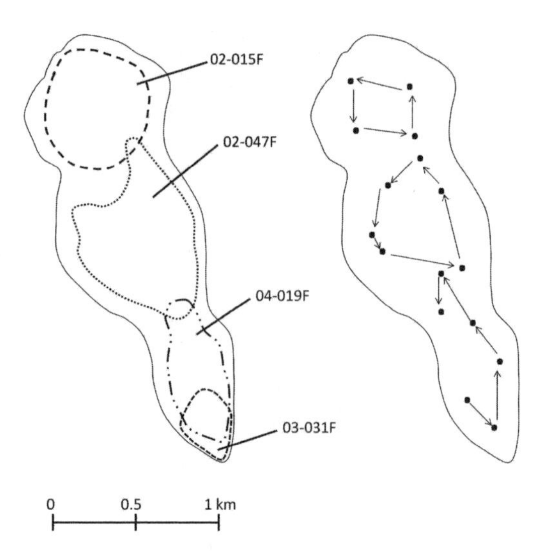

FIGURE 19. Territories of the four female black-footed ferrets monitored on the North Exclosure by the author over a summer season *(left)*. On the right is an illustration of the counterclockwise movement patterns of the three most commonly observed female ferrets as they changed the location of their dens from one prairie dog burrow to the next over the course of the summer litter-rearing season.

any one point in time. I made mental maps of how the three females had moved their young from burrow to burrow over the course of the summer. Each of them had moved around her territory in a counterclockwise direction, never going too far into the range of another female. They seemed to know if and when they were invading, and when they did, the next dens they occupied were back within their own core territory. In this way, each was stealing bits from the edges of the other's territory when her neighbor was not around.

. . .

By mid-August, the warm summer nights began to slowly lengthen ahead of the coming fall. The prairie dogs pups that had survived were so large that they were indistinguishable from their parents, and the maturation of ferret kits was nearly complete as they began to explore burrows that were separate from their central den burrow. They seemed to be testing their independence, at first separating from others only by fifty to a hundred feet during the night before coming back to the communal family den for sleep during the day. Later, they spent the entire day sleeping in a separate burrow away from the family. Kits were now the same size or more frequently larger than their mother, but were easily distinguished from her by their relatively plump bodies with fresh, clean coats. Mothers had been losing weight consistently since childbirth, first from the stress of lactation without feeding, and then the stress of killing prairie dogs with increasing frequency while having to share most of the meal with growing kits.

Compared to long June nights of little above ground activity, August was a time of near-constant action by the ferrets. The kits all came up together and stayed above ground for long periods of the night. They learned to poke their heads up upon first coming above ground and scanned the landscape for predators, then after a while, beginning with the most inquisitive kit, they would all come above ground to stretch their legs and play. This allowed their newly muscled, tubular bodies to move in a three-dimensional world not limited to the four-inch-diameter underground burrows. They played with each other in what Clark termed a "stiff-legged dance," lifting front legs into the air and stomping them down in unison, and then arching the back and doing the same with the rear legs in a bucking-bronco routine. The kits approached one another and then backed off quickly, mouths open, chattering at each other without clamping down. They bit at each other's tails and turned in quick circles. In Mellette County in the 1960s, Henderson observed a ferret executing a mid-air somersault.

Even solitary kits that had lost their siblings seemed to initiate random bits of play. A male kit that for the past two days had stayed in a burrow a hundred feet away from his mother, 04–019F, came above ground at dawn and attacked the reflective driveway marker I had left six feet from his burrow entrance. Noticing it as a foreign object on the flat prairie, like a flagpole in a mowed field, he reared back and made wild leaps toward the red disk on the top. Backing away, he lunged again and again, raising tufts of dust for twenty minutes. Then, as if surprised by his own tail on a particularly awkward landing, he scurried back to his burrow, only to turn and poke his head back out, repeating the attacks for ten more minutes.

It is when the kits were this playful and curious that I began to trap them. I would sit in my parked truck, trap and flashlight in hand, with spotlight fixed on them as they played. I would wait for one to isolate itself apart from the family. When one made such an adventurous trip and darted down an unfamiliar burrow, I would run out of the truck and push the trap into the burrow so that the ferret would have to emerge into the trap if it were to rejoin its family. Given that the young kits were not yet ready for full independence, with enough patience I usually could capture entire family groups in a couple of nights. Speed and timing were key. If I started trying to trap too early in the season when the kits were still together in a single burrow, I would usually only catch the same inquisitive family member. If I started trapping too late in the season, some of the kits would have dispersed before I was able to mark them and tell which mother was theirs. I also needed to be careful not to take too long in trapping one ferret when it was away from its family group. Keeping the kit away from its family group for a few hours or days increased the risk that the family group might move on, leaving the kit separated and potentially abandoned.

With the annual ferret roundup a week away, Travis came down to the basin early to help mark the young kits. Bill Perry had been pushed out, moved on, relocated, depending on whom you talk to. It's an all-too familiar situation for federal or state employees who advocate for black-footed ferret conservation, dating back to the Mellette County days with Bob Henderson, and even earlier. Regardless, Bill and his family had moved to Washington, D.C., and Travis had personally taken on the challenge of monitoring the ferret population of Conata Basin. He made up for the scaled-back role the U.S. Forest Service had decided to take on ferret conservation in Bill's absence by borrowing equipment and finding donors and volunteers to keep up the monitoring of the Conata population. He oversaw the entire effort from his small rental home in Wall, South Dakota, where he served as founder and sole employee of a nongovernment organization, Prairie Wildlife Research, the first and only group of its kind that was focused on conservation of the black-footed ferret.

Travis parked his purple Dodge pickup and camper trailer-turned-ferret-hospital just south of the North Exclosure. As I trapped the young ferrets and brought them to him, he began the process of recording them into the larger population. He placed them under light anesthesia as he implanted microchip tags under their skin, one at the shoulders and one at the rump, took measurements, gave vaccine doses, and assigned them their lifetime numbers: 05–001, 05–002, and so on. Before releasing them back into the wild, he applied black hair dye to their chests so I could visually identify them without approaching and disturbing them. The first kit was marked with a 1, the second kit with a 2, until the single digits were exhausted and we moved to letters like *B*, *V*, and then symbols like + and = . By the end of the first week of September, all ten kits and adults on the North Exclosure were marked, and Travis moved on to start monitoring and capturing kits in the rest of the basin.

. . .

For most of September, I was in a state of confusion as kits ran around constantly through the night, no longer paying attention to their mother's territorial boundaries. Young ferrets with die-marked chests ran between burrows and even crossed the entire colony in a night, movements unheard of a few weeks earlier. My maps and notes were a confusion of letters and numbers and GPS coordinates. V moving near = , in dome burrow just south of first den site of 02–047F. Male + along park boundary running north. Unmarked ferret on center of colony near 7.

By the date I had scheduled as my last night of spotlighting in early October, some sense of normality and order had returned because nearly all of the kits had dispersed off the North Exclosure and I began to see the same half-dozen ferrets every evening. The kits of female 02–015F all had left her and the colony for other parts of the basin. Female 02–047F had not been seen in four days, and because she was three years old and most wild ferrets do not live past the age of five, I feared the toll of raising four kits had been too much for her and that she had died. Or perhaps she had moved on to another part of the basin, making room for her kits? Regardless, in her absence it appeared that three of her kits had stayed in the North Exclosure, with 3 claiming the middle, 6 claiming the west, and 4 the east portion of her former territory.

On the lower end of the colony, female 04–019F, who had produced two kits, still had male kit V with her. Her other kit, 8, was also still on the colony but had been alone for nearly two weeks and I assumed she was capable of killing on her own. There were two additional male kits, 05–014M and 05–012M, who were very active above ground most of the night. Both had been born and marked by Travis on the nearby South Exclosure prairie dog colony, and they crossed the narrow gully to reach the North Exclosure.

Remarkably, the most commonly observed and wide-ranging male throughout the summer, 03–034M, had been missing for the past two weeks. In his absence, the two new males from the South Exclosure explored and scent marked over large portions of the middle and lower parts of the colony. As if on cue, the two resident males, 04–006M and 04–003M, also became increasingly active. But even in the apparent absence of 03–034M, these two resident males were hesitant to invade his former territory, sticking to the northern two-thirds of the colony where they had overlapped each other.

As the cool fall nights set in, the cycling of male ferrets through the colony became unpredictable as they seemed to show up randomly on certain nights, likely trying to claim territories while the females remained secretive below ground. As I prepared to leave the basin for the year, I was already curious about what I would find the next spring. Who would make it through the long, harsh winter? Would a single male take over much of the colony? Which females would raise kits, and would the mothers continue to tolerate their young being in their territories or force them out? Would reproduction be as high for the new females as it was for their older mothers?

. . .

My last night with the ferrets was cut short by a 2:00 A.M. thunderstorm. A front rolled in from the west, blocking the light of a full moon that had been creating shadows on the badlands walls. To avoid getting trapped on my last night, I high-tailed it out of the North Exclosure. Making my way toward the highway, I splashed through the first and second ditches that were already flowing with a foot of water. The prairie turned to six inches of clay mud as I fishtailed the truck back and forth to make it the last three hundred feet and onto the pavement of Highway 44. I maxed out the windshield wipers as rain washed the clods of kicked-up prairie from my hood and fenders.

I had not yet picked up speed when at mile marker 27, two men ran out from under a lone cottonwood tree and into the beam of my headlights to wave me down. I hesitated for a moment, asking myself who in their right mind would be out here in the middle of the prairie so far from human habitation. I passed them, saw the desperation on their dark faces, and then began to slow. Pulling to the side of the road, I watched them trot after the truck in the rearview mirror. I took a quick survey of the truck for a weapon, a knife at least, and only found the large flashlight on the floorboards. When the men entered through the passenger's door, they were drenched and smelled of the wet prairie earth I had just plowed through on my way to the asphalt. They slid onto the beige vinyl bench seat beside me and wiped the water from their faces. They were giants with pock-marked skin, braided black ponytails, and enormous hands that would have trouble fitting into the trigger guard of a pistol but could wring a person's neck with a quick snap rather than a prolonged strangle.

I asked them where they were headed. They mumbled and I finally said that I was headed toward Interior and that I could drop them off there. They didn't object, so we drove in silence for ten miles as the rain picked up even more and turned the road to a thin sheen of black water. The radio played a South Dakota news story of a couple of girls from the Pine Ridge Reservation. After a basketball game between a reservation team and the local town, the reservation girls drove themselves home. Stopping for ice cream at an off-the-reservation drive-in on the way home, where local high school farm boys parked and talked into the night about girls, or pranks, or graduation and signing up for the National Guard. The radio voice told us that as the girls pulled away the boys yelled insults at the girls that made me cringe next to my two hitchhikers. The girls drove on, boldly turned around the block, and drove back by the drive-in. This time they chucked their ice cream cones, which landed squarely on a boy's truck windshield as they sped away. The boys jumped into their pickup and gunned the engine, catching the small sedan outside of town on the two-lane road headed south toward the reservation. The pickup pulled alongside the sedan in the lane for oncoming traffic and a boy in the passenger's seat pulled a shotgun and smiled. The girls screamed and drove on. No one was shot, but the story made national news.

I looked over at the two giants beside me and the one closest to me asked for water, despite being soaking wet and the rain pouring outside. I lied and said I had none.

"What are you guys doing all the way out here?" I asked.

"Walking home from my cousin's place," the one next to me replied.

"Where's that?"

"Scenic." Twelve miles to the west, they still had at least ten miles to go to get to the nearest town.

I rolled down my window a crack and the wind chilled the air inside the cab. Despite their bulk, the men huddled over and began to shiver as I pulled into Interior.

"I'm heading north from here, through the park." I paused only to catch my breath and not allow them enough time to ask to continue on with me past this last patch of human habitation. "Where do you want to be let off?"

They pointed just off the highway past the center of town, and I pulled over next to an abandoned home with a barbed wire fence around it behind the Horseshoe Bar. They got out of the truck into the rain without thanking me and climbed over the perimeter fence. Once over, they headed at a dead run into the darkness and toward an old-fashioned well. They stopped and began to feverishly pump the handle for water.

I drove away and the rain continued to pour down as they finished drinking and disappeared into the shadows of the gutted homestead. I took the road that snaked up through the badlands and over the wall toward home, thinking it odd that they never asked what I was doing out here late at night, knowing they could see my spotlight from where they had waited under the cottonwood tree by the highway. I wondered if they knew this land as well as I was starting to know it: the cycles of the moon, the coyotes and badgers teaming together, the six-inch tiger salamanders coming above ground on the prairie following torrential downpours such as this. Did they know what a black-footed ferret looked like? Had they ever taken long walks through this area as young men on vision quests as their Sioux forefathers once did? Were they truly just walking from town to town when the storm rolled in, or were they just camped under that tree after a multiday meth high until awakened by the rain. On the other side of the badlands wall, the rain had stopped and windows fogged with heat and moisture that radiated up from the vinyl bench seat beside me. I rolled the window down and the air circulated the scents they left of musk, mud, and chemicals.

. . .

The following year, I was able to wrangle some money from wildlife agencies and foundations to expand the research project by hiring two spotlighters to help me keep track of two ferret populations simultaneously, one at the North Exclosure and one at UL Bend. I told the donors that by monitoring both populations of ferrets simultaneously, I could directly compare how ferrets at Conata matched up with those at UL Bend. I knew both populations tended to breed at the same time and give birth at the same time, but I didn't know why recruitment (the ability of females to produce surviving young animals each year to increase or maintain the population overall) was lower at UL Bend than at the Conata Basin. Why, when, and how did this disparity come about? Perhaps the UL Bend population was struggling because females needed to travel more to rear their kits on colonies where prairie dogs and their burrows were fewer and farther between. By comparing female ferret behavior simultaneously between these two populations, I expected to begin to answer these subtle, but important, questions.

At the North Exclosure, we started spotlighting in June and found that male ferret 04–013M had remained along with male kit 05–014M from last fall. Five females remained, including 04–019F and 03–031F from the previous year. One of 04–019F's female kits also had survived the winter and remained near her mother's territory on the lower end of the colony. Females 02–015F and 02–047F, who had been present the year before, had either moved on or, more likely for three-year-olds, didn't make it through the winter. Poetically, a female from each of their litters the year before had remained. Female 02–047F's kit 05–003F occupied a territory just north of where she had been born and raised. Female 02–015F's kit 05–005F occupied a territory almost identical to the one her mother had previously used.

This meant that at least three of the ten kits born and raised to dispersal age on the colony the previous year had survived to adulthood, a 30 percent kit survival rate at a minimum. Assuming one or two of the other kits survived and dispersed to other parts of the basin, overwinter survival was on par with what biologists at the University of Wyoming predicted for this population based on Travis's long-term records. They had found that on average, female ferret kits had a 57 percent chance of surviving to their first birthdays. And by their first birthdays, they should already be pregnant or have just given birth to their own litters of kits, of which an average of 1.5 to 2 kits would survive to dispersal age. It was the type of quick turnaround needed for an animal with a three- to five-year life expectancy.

In truth, the 57 percent survival of kits to adulthood the following summer estimated by the University of Wyoming group was optimistic. Studies of another large and self-sustaining ferret population in Shirley Basin, Wyoming, suggested that survival of kits to their first birthdays is more like 39 percent. Further, surviving that first year to adulthood seems to be the critical gauntlet because the probability of a female ferret surviving from her first birthday to her second birthday is twice as high as that first year. This low first-year survival likely explains why, during the annual September ferret roundups at reintroduction sites, a majority of ferrets encountered are young-of-the-year animals (approximately 65 percent), while there is a much rarer chance of seeing a ferret that is one (19 percent), two (9 percent), three (4 percent), or four years old (2 percent).

Remarkably, regardless of age, female productivity remains constant throughout their short lives. This tells us that there is little benefit in being older and potentially wiser. In contrast to long-lived species that have a chance to improve their skill of rearing young to adulthood, scientists currently posit that ferrets show no distinct learning curve for how to successfully raise a large litter. Once a female ferret is born, the clock is ticking, and she must be a good mother in the first year, because there is only a marginal chance she will survive to age two. The overall take-home message is that with such a short life span and limited time period of producing young, females must have as many kits as possible as quickly as possible. Further, the ability of a female to produce at least a couple of kits in her first year that last through the winter likely dictates the success or failure of a small population, particularly one like UL Bend.

At UL Bend, Randy Matchett had already documented this problem through his annual survey results. He had shown that litter production at UL Bend was low compared to Conata. Yet it was unclear whether ferrets at UL Bend were just giving birth to fewer young, or whether litters were dwindling in number prior to the time when researchers typically see them above ground in August. How large were the UL Bend litters when they were born? I was able to get late June and early July litter counts for most females on North Exclosure the previous year, but no one had seen a female move her kits in early June when they were still less than a quarter of their mother's weight. Given that females rarely bring their kits above ground except for the briefest of periods during that time of year, I decided to go subterranean. I bought a plumber's video camera, composed of a small golf-ball-sized camera on the end of a thirty-foot cable that attached to a battery-powered TV monitor. My goal was to run the camera down the burrow and count the kits in the litter while they were in their den burrow.

There were few accounts of how far below ground the den chambers of ferrets are and what they look like. During the 1960s, Robert Sheets, then a graduate student at South Dakota State University, excavated ferret dens in Mellette County, South Dakota, for his thesis research. He used a backhoe to dig deep into the prairie dog burrows he knew ferrets had used, and then went in by shovel and finally by hand to map the ferret dens and collect scat samples from them. Typically they were more than fourteen feet below ground and beyond the reach of the backhoe. Complex mazes of burrow systems extended up to sixty-two feet in length and occasionally had vertical descents of seven to ten feet before making abrupt right-angle turns. To help thread my camera through the burrows, I bought a plastic toy truck with oversized tires, cut off the cab, and I tied the camera onto the frame of the truck to allow the cable camera to roll on the ground and hopefully navigate sharp corners. When the delicate wheels failed, I planned to use brute force to push and twist the camera around turns, so I fortified the narrow cable with a spool of quarter-inch flexible plumber's snake.

Prior to disturbing a female and her litter, I tested the camera on male ferret 04–013M, who had been holed up in a burrow alone for the past three days. I turned on the camera, placed the modified toy truck inside the burrow, and began threading the camera straight down. Not five feet below ground, the male was in view. He no doubt heard the commotion of the wheels and the scrape of the cable on burrow walls, and I watched on the black-and-white screen as he approached the infrared camera. He smelled its lens, distorting his face in my screen view, then retreated back down the burrow. I tried to follow him deeper with the camera. The burrow continued straight down, opened into a one-square-foot chamber and then narrowed again to a four-inch-diameter burrow and continued farther straight down. At nine feet below ground, the burrow took a hard right turn and the toy truck smashed head-on into the soil, the camera lens bogged down, and the cable bent under my pressure. I pushed and pulled to try to turn the truck, twisting the cable to make the tires grip on the sides of the burrow and slide the camera to the right. After ten minutes of violent twisting and pushing, it finally turned, went one foot, and the male was again in view and checking the camera head-on. This time he opened his mouth and I could hear the small chattering sounds reverberate up the burrow. He was mad, perhaps thinking it was a strange ungraceful snake. He made a false charge, nipping at the lens. What at first was a curiosity to him now was a threat.

"And this was only a male," I thought to myself. No litter to protect and much bigger than a female. A female ferret would surely be more aggressive in her defense of the burrow against an intruding camera. This would cause more stress to an already physiologically maxed-out female. I remembered the captive-breeding scares of females abandoning their kits, even cannibalizing them. I thought of the dangerous balance between protecting her kits and protecting herself in her short-lived world surrounded by predators. If I scared her from her kits and she associated the burrow with a threat, there was the risk that she might not return to nurse them. The potential benefit of ever getting to a female with the camera and counting her young kits was surely negated by the amount of disturbance I would create. I decided to give up my cable camera.

With plans of counting kits underground thwarted, I turned to plan B: to try to count during the brief moments when mothers move their kits between dens above ground. To accomplish this would require almost constant monitoring of active dens, a monumental feat of patience that neither my assistant, Dave, nor I could perform. So again I went high-tech and borrowed technology from an ornithologist friend who used time-lapse video cameras to monitor the nests of songbirds in the Midwest. From a security company in Texas, I ordered a time-lapse video recorder of the type used in police cars that runs on a 12-volt car battery. I also purchased a small infrared home security video camera that could be placed outside the burrow entrance and not disturb the ferrets, a mass of cables and connectors, and a marine battery to run the equipment through the night. To hide the apparatus while also protecting it from the elements and the trampling hooves or prying noses of cattle, I purchased an igloo-shaped, plastic dog house and painted it buff brown like the prairie. I drilled a hole in the back wall for the camera to stick out so I that I could use the normal igloo entrance to reach in to play with the cables and cassettes.

To test out my new contraption, I placed the camera hidden in the dog igloo facing a burrow at which I had seen 05–003F the night before. I put in a blank VHS tape and let it run for the next twenty hours, and then put in a new tape before returning home to watch the footage. At home, I sat down in the mornings after a night of spotlighting and watched the black-and-white image of a rounded burrow opening, looking for any flash of movement that would indicate that the female ferret had popped up out of the burrow.

After the third night without movement, I worried that 05–003F had used an adjoining, backdoor burrow entrance to sneak her kits out. Perhaps I had missed her leaving in the ten hours that elapsed from when I last saw her at the burrow to when I installed the camera. Or worse, perhaps my dog igloo camera scared her to the point where she was hesitant to leave, put off by the new, round object placed fifteen feet away. Even though it was painted to match the short dry grasses, it was in plain view on the flat prairie. Perhaps she thought it was a small hill providing cover for a coyote that lay in wait for her.

Finally, on the video of the fourth night, I saw movement. First, just her head, then her body, then she moved out of the frame away from the burrow. Ten minutes passed and she returned, stopping at the burrow entrance to glance at the dog igloo before going underground. Within two minutes, she was back in frame, poking her head above ground to periscope in all directions for the presence of predators. She paused briefly to look at the camera, and then ducked back down just inside the burrow. Two seconds later, she emerged with a kit in her mouth. Its eyes were still closed and it was only one-fifth the size of its mother. She picked up the little hot-dog-sized and -shaped kit just behind the head, at the nape of its neck. The kit was unresponsive, in a slack position looking dead, but more likely genetically programmed not to fight or fuss. The kit's fur was short and pale, with only the black mask and feet of its mother. The black-tipped tail was less than an inch long and tucked in tight to the body. With neck outstretched the mother was able to keep the kit dangling above the ground, pausing for an instant before she quickly darted out of view.

After three minutes, she returned from the same direction she left, taking the quickest and shortest path between the old and new den burrows to avoid catching the eye of a predator. She repeated this pattern for two more kits, settling the family into the new burrow where she had stashed a freshly killed prairie dog. After the final kit, she never returned to view, telling me that her mid-June litter consisted of three kits. Why finally abandon that den for another? What told her this was no longer a good site? It seemed to me that there was far greater risk in moving to a new, unknown den where prairie dogs or rattlesnakes might reside and harm the kits or their mother. Maybe she moved from the old den site because it was ripe with the smell of scat and a decaying prairie dog carcass, potentially attracting a badger to excavate and investigate the smell, or causing a prairie dog to block the burrow entrance. Regardless, she was done with it now, perhaps remembering its location only as a good site for possible use next year.

. . .

After a summer of using the dog igloo camera, I learned that mothers varied in how long they waited to change den burrows. In June, when the kits were young, a mother would go off to kill a prairie dog only every five or six days, after which she would take the freshly killed prey to a new burrow with an appropriate den chamber and move the kits to this burrow for the meal. The burrow, located between the kill site and the old den burrow, Con Hillman termed a "cache burrow." This system of hunting and moving kits that I saw was also spotted in the last wild populations by Con in Mellette County and by Ray Paunovich in Meeteetse. Further, because mothers had to kill more frequently as the kits matured, they had to change dens more often over the course of the summer. By August, females were moving kits to new den burrows every one to three days.

When I started to map ferret locations over the course of the second year at Conata, I found that the two female kits born the previous year on the North Exclosure had largely taken over the territories vacated by their mothers. On the territory at the north end of the colony vacated by 02–015F, her daughter 05–005F produced three kits. Female

04–019F remained in her previous territory, producing one kit. Female 04–019F tolerated the presence of her daughter, 05–008F, within the territory where she had been raised the previous year. The daughter produced five kits, two of which made it to dispersal age. Female 03–031F also remained on the lower end of the colony as she had done the previous year, overlapping with 04–019F and raising a whopping five kits to dispersal age. Given her lack of production the previous year, I wondered why the sudden turnaround. She was certainly a good mother, so I chalked her failure to produce the previous year to a chance encounter with a predator. Perhaps losing her kits had the benefit of leaving her body in better condition to make it through the winter and explained why she was able to raise such a large litter the following year.

Overall kit production was high, and the lower half of the North Exclosure colony in particular had produced three litters and a total of eight kits, a large number for such a small area. By contrast, the three females we were able to monitor at UL Bend produced only seven kits. Although this difference was small on a per-ferret basis, there was a bigger issue at play. Unlike female ferrets at Conata Basin, females at UL Bend did not use large portions of the available habitat. Entire colonies of prairie dogs were without ferrets. Was it that ferrets were not finding these areas, or did they find the distance and exposure of moving between colonies too risky? More likely, ferrets were colonizing these isolated patches of habitat but simply not surviving on them. Biologists released ferrets directly onto these patches over the years at UL Bend, yet they do not persist there today. Was it because prairie dog populations were so low at portions of UL Bend that they simply could not sustain the food needs of a female ferret and her litter? Perhaps the ferret families were literally eating themselves out of house and home.

We knew that the density of prairie dogs was key to how many ferrets could occupy an area. My monitoring on the North Exclosure showed that female ferrets seemed to use smaller areas and be more likely to tolerate overlapping with other females when prairie dogs were abundant. By contrast, at UL Bend, where prairie dog densities were much lower, female ferrets had larger territories and didn't tolerate as much territory overlap. This meant that if a site had a high density of prairie dogs and their burrows, more ferrets could occupy a given area. On North Exclosure, females seemed to go so far as to share areas where their offspring and other ferrets could rear kits. At a larger scale, it was also likely that these areas with high densities of prairie dogs were important sources of ferrets for the rest of the basin, providing highly productive areas from which ferrets could disperse to less productive areas and eke out a living—a sort of seed stock with which to populate the larger Conata Basin with ferrets each year. But what about sites like UL Bend, with only a few, small, highly productive areas? Could a few females keep a population going? What happened to those individuals in the first year of life, which was so critical to maintaining a ferret population, when such patches were few and far between? Contrasting the success of Conata Basin to UL Bend seemed to suggest that size of the prairie dog colony matters. But was it size alone? What about El Cuervo, where despite the immense size of prairie dog colonies, their density seemed to be too low to sustain ferrets?

Back at the office, I tested these theories with math. I compiled statistics from each of the eleven reintroduction sites that had at least five years of attempted reintroductions, ideally giving ferrets enough time to catch on. I formed and tested competing hypotheses that the success of ferret populations was related to the number of ferrets that were released, or to the time that has been allowed for them to take off. I also tested hypotheses that prairie dog species, colony size, or colony density was a significant driver of success. And, finally, I tested a hypothesis that the number of prairie dogs, the raw biomass of prey in terms of both the size and density of prairie dogs across the entire reintroduction site, was key. After crunching the numbers, the results were clear. My models suggested that no sites with prairie dog colonies smaller than ten thousand acres in size were likely to be successful, but as seen in El Cuervo, it was more complex than that. The most important metric for predicting ferret reintroduction success was prairie dog biomass, meaning that the successful establishment of ferret populations was tied to an interplay of prairie dog colony size and density. This suggested that increasing prairie dog biomass on reintroduction sites, many of which were smaller than four thousand acres in size, was likely to be the answer to many of our problems. The finding pointed a finger at the persecution of prairie dogs, but further stirred the pot because of the revelation that such large numbers of prairie dogs were needed. Numbers of prairie dogs far beyond minimum energetic demands of ferrets on which we previously based site evaluation forced us to expand our thinking beyond moderate increases in prairie dogs through translocations and habitat improvement, to larger-scale prairie dog conservation. This was hard evidence that landowners, biologists, politicians, and the rest of

America needed to envision prairie dog conservation at a landscape scale, like what Bill Perry had created at Conata Basin.

. . .

After following the ferrets of North Exclosure through dispersal into late September for a second year, I took a night off to have drinks with Doug Albertson, whom I hadn't seen since Mexico. Doug was from the town of Belle Fourche on the other side of the Black Hills, and despite quitting his job as head biologist at Badlands National Park and being unemployed and traveling for more than a year, he still kept his small home on a ranch south of Interior on the reservation. He lived simply and cheaply, remarkably without a satellite dish and television despite having the means to acquire them—a tough feat for a single man in such wide-open country.

We walked into the Horseshoe Bar in Interior, and there was only a handful of people in booths along the outer walls. Doug went straight to the stools by the bar and pulled out his wallet. I sat next to him and he ordered a Budweiser, and then one for me and the girl sitting two seats down. The etiquette seemed to be that if you went to the trouble of reaching for your wallet, you should buy everyone a drink, with the next round on them.

The girl thanked Doug, and he introduced me to her. Vanessa was a PhD student working in the park, studying the bighorn sheep population. Doug told me she lived in a trailer parked on the ranch of Charlie Kruse, a vocal individual who we all knew was leading a group of area ranchers in petitioning the Forest Service to control prairie dogs in the basin.

"You feel safe living out there?" I asked her, meaning as a biologist who worked to conserve wildlife more than as a single woman all alone.

"Oh, Charlie is a nice guy, as long as you don't work on ferrets or prairie dogs." She looked up at me with a grin, knowing I was one of the folks working in the basin.

"Besides, I always keep my Glock and hollow points nearby," she continued.

I looked for a hint of a smile, showing she was kidding, but found none. I suddenly remembered that I had first met Vanessa a few years earlier when a group of us was helping the park decide how best to manage and thin out the growing bison population. She acted as a tour guide back then and was dressed much differently, layered in the local standard midwinter outfit of canvas coveralls. With a large wad of tobacco under her lip, I had no doubt that she was well armed.

"You had a good deer season, I hear," Doug chimed in to fill the silence.

"Filled the freezer."

Doug turned to me.

"She filled her tags on the first day. Hiking in on the opening day, killing a mule deer, quartering it up, humping it three miles back to her truck. Hanging it at home, then going back up to get her second deer. A freezer full before dinner."

But tonight she was dressed in tight jeans and a small, maroon corset, with bulging pale skin and auburn hair done up in curls, lips painted bright red.

"You are out of South Dakota State, right?" I asked.

"Yep, Harvard of the plains that school is. Better than any East Coast or Midwest education," she said while looking up from her drink to see me squirm, knowing I was spending the winters in Missouri to get my degree. I hid any annoyance and laughed, trying to focus my eyes on her without staring at her eyes to give any sign of discomfort. I tried to stare at the only unprovoking part of her, her ears, fighting the urge to look away or down and inadvertently into the prodigious cleft of cleavage created by her corset. She was satisfied, or perhaps not, but continued on.

"I plan on finishing next year."

"What then?"

"Don't know. Tonight I'm just lookin' to get knocked up."

"You still with John?" Doug asked.

"Yeah, maybe," she said turning to look back at the corner of the bar where two guys and a girl were laughing and getting drunk in a booth. The girl was dark-skinned and young, too young for the oily, thin man wearing a black T-shirt with cut-off sleeves. His bare arm was over both of her shoulders as they sat on the bench seat and he laughed. The other man sat opposite them, tall and round, with a white T-shirt and red ballcap advertising the latest genetically modified wheat product from an international conglomerate grain or seed company. The agricultural co-op kept farmers well stocked in free advertising.

Vanessa spoke up, "We'll see if he stays sober enough tonight."

We finished our beer and ordered another round. A few cowboys came into the bar and started the juke box in the corner. Another walked in and put his arms around Vanessa from behind. By ten o'clock, things had picked up. More men arrived, all dressed in their daytime work clothes. Cowboy boots thudded on the old floor, a pool table in the back was uncovered, and fluorescent lights were turned on.

At midnight, two teenage boys walked in wearing clean, white T-shirts and gold necklace chains, black, baggy jean shorts worn low, and tennis shoes. They looked more at home at a dance video set than at this dusty bar with dollar bills signed and stuck to the ceiling with bubble gum, staples, tape, anything that was handy.

Doug saw me looking them over. "Who are they?"

"Oh, just two kids from the reservation," Doug replied.

"They must have money."

"Yeah, their family is one of the top beer sellers on the reservation."

He saw me look confused, seeing that I knew alcohol was illegal on the reservation.

"Importers. Bootleggers," he continued. Whole families worked in mafia-style collusion and specialized in bringing alcohol illegally onto the dry reservation. The Pine Ridge Indian Reservation occupies the majority of the second-poorest county in America, with a per capita income of $6,286 and more than 90 percent of its residents living in poverty. Concerned tribal leaders blame many of their problems on alcohol, with more than a quarter of children born with fetal alcohol disorders and an average life expectancy of less than fifty years. But bootlegging business was big and booming. Beer was sold from the back of bars outside reservation boundaries by the truckload. Authorities were paid off. Supply fulfilling demand.

South of the reservation, just over the border in Nebraska, the town of Whiteclay did brisk Budweiser business, selling more than 4.6 million cans of beer and malt liquor a year. With the town of Whiteclay having a total population of fourteen, and only sixty-two people living in the entire zip code, that easily made their per capita beer sales the highest in the world. The true consumers who entered the gigantic beer coolers in one of the four booze shops in Whiteclay were Pine Ridge residents. Because alcohol was illegal just on the other side of the state and reservation line, men and women slept outside liquor stores and camped in the fields outside of town. Earlier that year, Army veteran Bryan Blue Bird was sleeping off the night before when the county unknowingly set a prescribed fire in the fields outside of Whiteclay. He was still in the hospital after receiving burns to 25 percent of his body.

We left the Horseshoe and headed over to the other bar in town, the Wagon Wheel, which sat a little farther off the highway. The bar-to-church ratio was two to one, and there was no gas station or food store to speak of. At the Wagon Wheel, there was a sign taped to the door advertising that country singer John Michael Montgomery was coming to town to perform in a benefit concert on Circle View Ranch just west of town. The benefit was for the group of Interior cattle ranchers led by Charlie Kruse who had called for land management change in the Conata Basin that would undo the years of work Bill Perry put in to creating a protected area of ferrets. Their contention was that for the health of the grasslands and their cattle businesses, the Forest Service needed to go in and poison prairie dogs in the basin.

Beyond country music one-hit-wonders, Kruse's group was courting politicians and threatening lawsuits, the two things to which federal agencies really listen. Newly elected U.S. Senator John Thune was on their side:

> I am pleased that the Forest Service has responded to my concerns about the environmental destruction being caused by the overpopulation of prairie dogs on our grasslands. This amendment process is crucial to giving the Forest Service the opportunity for further analysis and broader management tools to protect the soil and vegetation, which are being decimated by prairie dogs in areas like the Conata Basin. Ranchers, land-owners and all South Dakotans deserve a healthy grasslands that is managed in an environmentally sound and responsible manner beneficial for cattle, wildlife, neighboring landowners and communities. Hopefully this process will allow the Forest Service to accomplish that goal.

Tickets were $50.

"Too rich for my blood," Doug said as he saw me stopping to read the flyer.

Inside the bar, the sparse crowd was comparatively older, drunker, and looking like they hadn't left the place in a couple of days. Neon lights outside brought in more light than the dim lights inside. Luckily, the bartender was cute and mercifully younger than the clientele, wearing a tight-fitting tan leather vest that showed off her dark skin and long black hair. By the look on her face when Doug stepped to the bar, she knew him very well. She smiled and asked when he got back into town, and why he hadn't called or stopped by.

We got our bottles of beer and found a table away from the crowded bar. I tried to continue our conversation about the door poster, knowing he wouldn't want to it to be overheard by anyone else in the place. In South Dakota, hatred of prairie dogs still divided the community as easily as religion or politics did in other parts of the country. Many locals seemed to keep the mind-set of European pioneers—that the Great Plains are something to be tolerated at worst, or conquered at best, a vast "cattle country" where places like the Conata Basin are just degraded cow pastures.

But Doug was different, and I knew he had a soft spot in his heart for ferrets and prairie dogs. He knew that to black-footed ferret biologists, the Conata Basin was Mecca, the first place in the world where a self-sustaining ferret population had been restored. A success in an otherwise depressing story of twenty years of attempts to release captive-raised ferrets onto nineteen locations spread from Canada to Mexico.

"The Basin is a damn mess," I started with.

"Yep."

"You know they are calling for poisoning it down to nothing?" I said a little too loudly. The alcohol was beginning to remove my social graces and allowed brewing frustration to come forward after two years of solitary time spent spotlighting just down the road.

"Forest Service is going to have a hard time not poisoning some boundaries," Doug reasoned. I nodded and took a deep drink. We both knew it was true because poisoning had already occurred to some degree on the edges of the Forest Service land. Forest Service staff from Wall came down into the basin for a few days and surveyed for ferrets before contractors arrived to put down bait to kill prairie dogs on public lands. State-funded prairie dog poisoning experts were hired to kill all prairie dogs on public land within a half-mile-wide buffer strip inside of the fence line with private ranchers.

This perennial killing each fall resulted in empty burrows that were ripe to be recolonized in the spring by hungry prairie dogs dispersing from the center of the basin. A cycle of prairie dog conflict that meant guaranteed income security to poisoners and a net drain on the prairie dog population. To make matters worse, a drought was driving prairie dogs to expand their range in the basin or risk starvation. Studies from across the West showed that both the density of prairie dogs in an area and their ability to survive and reproduce varied with annual changes in rainfall and the availability of forage. So during the most recent drought, although many assumed the expansion of burrows into new areas meant that prairie dog population was growing like crazy within the basin and efforts to cull them back needed to be stepped up, in reality their densities and population size were decreasing.

"The new alternative proposed to the Forest Service by politicians is to shrink the prairie dog colonies down to eight thousand acres," I continued. The juke box was turned up by someone in the corner, and we were silent. There were no solutions, no easy fixes. I thought of the ranchers in El Cuervo and their tolerance of prairie dogs even though they were in far more dire ecological and economic straits. They realized what scientists and economists had proven for years: that the overall cost of poisoning prairie dogs did not balance the benefit to cattle weight gain. Without government subsidies, there was no reason to poison. Somehow, in layers of Western mythos, the U.S. taxpayers had overlooked that in allowing their tax dollars to be spent on prairie dog poisoning, it will not lower the price of beef at their local grocery store. There is no logic to the large-scale eradication, only prejudice and paranoia that has been passed down through the generations.

"Might have to move some ferrets up into the park," Doug said with a smile, trying to halt my building depression, all along knowing that if the park habitat were that good, ferrets already would have moved there in force. He and I both knew that Conata Basin was the only hope ferrets had in this wide swath of mixed-grass prairie.

"At that size, the basin will be like Montana all over again."

CHAPTER 11

Plague

At dawn on the eighteenth of June, it was already 80 degrees and likely to get over 100 degrees by noon. To beat the heat, we gathered at the U.S. Forest Service office parking lot each morning at six. Carl, a federal biologist stationed in Wall, and I waited until twenty after, and left without Jenny.

Jenny had just graduated from college and moved west to live with her dad, who was a trapper for the State of South Dakota. She worked with us as a duster during the day, and in the evenings drew all the young men to her in Wall with stories of Illinois sorority parties and three types of outfits for "going out." She was the fresh face in a small town along the interstate where stout young ranchers came to one of the two bars in the evenings looking for love, but available young women usually stopped only for the night on their way to somewhere else.

Carl and I drove south over the badlands wall, through the park and into the Conata Basin where it had rained overnight for the first time in three weeks. After three years spent in the basin, I had learned how a summer rain after a dry spell smelled different than other rains. Light dust particles were puffed up into the air by the impact of raindrops and then weighted back down to nose level with morning dew. The storm clouds were now headed farther south over the Pine Ridge Reservation and morning light was pouring through the cloudless sky and reaching the spires, pinnacles, and walls of badlands lining the basin. Normally muted brown sedimentary layers were brought into contrast

FIGURE 20. Aerial image of the badlands formations within Badlands National Park, near Interior, South Dakota.

by the wetness, turning some layers into salt-white veins against an indigo blue sky.

We turned off the highway at milepost 14 where there was a break in the three-strand barbed wire fence. Fresh tire tracks told us someone had already been down our road this morning, and led to a rusted GMC camper van that was parked in the middle of the two-track road just in from the highway. I leaned on the horn and Jenny climbed out of the backdoor in her "going out" skin tight jeans and painted nails. The tight curls she had ironed into her hair had started to go flat and she walked to us barefoot, with heels and a small duffle bag of work clothes in hand.

"And she said she was a virgin," Carl said to me.

"Sometimes that is just something people say."

"Well, she yelled it to the entire bar last Friday night."

We shook our heads at the concept of our princess losing her twenty-two years of innocence to a rodeo boy from down on the reservation who wore a mohawk haircut under a Stetson hat. But we were also proud that she had the resolve to show up for work in the morning while clearly nursing a hangover.

Without a word, Jenny slunk into the back seat of the pickup and we continued around the camper van and into the basin. We followed the fence line north over open prairie toward the North Exclosure at the base of the badlands wall. Jenny changed into work clothes and boots in the back seat as we crossed a gulley and parked next to a trailer used as our field camp.

. . .

We were here ahead of the plague outbreaks that others had been battling in Montana for more than a decade and that had suddenly crossed over the theoretical plague line on the 102nd meridian. West of the meridian, plague had decimated ferret reintroduction sites in Colorado, Wyoming, and Montana. The last known wild ferret population in Meeteetse was under threat of a plague epizootic in 1985 when biologists began to spray a pesticide, Carbaryl, into eighty thousand prairie dog burrows to kill fleas that spread plague. Initially only 20 percent of the colony was lost, but much of the remaining population was lost in subsequent years after ferrets were removed from the site and brought to Sybille. Similarly, in southern Wyoming, the first ferret reintroduction site in Shirley Basin underwent plague epizootic outbreaks from 1991 to 1996 that reduced prairie dog populations from an area that covered thirty thousand acres to an area less than ten thousand acres in size, and the ferret population declined to five individuals. In central Montana, within weeks of releasing thirty-three ferrets onto Fort Belknap Indian Reservation in 1999, a plague epizootic ran through the prairie dog colony, reducing habitat by half and extirpating the reintroduced ferret population. Southwest of Fort Belknap, after four years of ferret reintroduction attempts on the 40-Complex north of UL Bend, a plague epizootic hit the prairie dog population. Treatment of the colony with Deltamethrin flea powder enabled six ferrets to survive, but the next year, when no flea powder was applied, the remaining prairie dog and ferret populations were extirpated.

After reaching the 102nd meridian by the 1980s, the disease seemed to stop its progress west for two decades. Then there was a report in 2004 of it appearing in Custer County, South Dakota, on the west side of the Black Hills just over the Wyoming–South Dakota line. By late July 2005, plague was confirmed in dead prairie dogs found on Pine Ridge Reservation in Shannon County, just south of the Conata Basin. In response to this threat, managers organized a massive prairie dog burrow-dusting campaign in the summer of 2005 in the basin, using the

FIGURE 21. Sign on the road leading into the Conata Basin warning visitors about the ongoing plague outbreak in the prairie dog population.

same technique pioneered twenty years earlier in Meeteetse. Biologists with a half-dozen federal agencies from across the country flooded into the basin and were outfitted with a four-wheeler and mechanized sprayer used to dispense Deltamethrin flea powder. Their goal was to inject five grams of the fine white flea-killing dust into each burrow in prairie dog colonies extending over a large region of the basin in an attempt to protect the prairie dogs and ferrets from plague jumping north out of the reservation.

Luckily, the 2005 treatment worked and plague did not come north. The next year, dusting was repeated and there were again no reports of plague jumping into the area. By 2007 managers thought the disease risk had slackened. There was little money to keep pouring into the basin to keep up the dusting effort. It was down to just the three of us.

Every day we returned to the basin to dispense flea powder that was purchased by the ton but applied carefully by the thimbleful by stopping at a burrow for ten seconds, injecting five grams of flea powder deep into the burrow, and then moving on to the next one. We developed a pattern of side-by-side teamwork east to west, back and forth across the prairie. We began to pay attention only to clicking off of how many burrows we had visited, losing track of other numbers associated with hours, days, and weeks. We became used to the burn of the sun on our skin mixed with a layer of white pesticide and badlands dust. We kept our eyes to the west for potential storms and squinted to look for the far fence line that would mean we had finished for the year.

At the end of each day we would get back at dusk, with energy only enough to scrounge for food at the simple grocery store in town. After defrosting a burrito in the hotel microwave, I struggled with the need to eat against the urge to collapse into bed. Undressing in the evening in front of the mirror, I saw tan lines on my arms that made me look like two different people. There was a ring of white on my forehead from my straw cowboy hat. My hands had turned black from the rubber grips of the four-wheeler. I felt the satisfaction of a day's work on the Wyoming ranch of my childhood. I felt primal, like a long-distance hiker, thin, with ribs showing and a manic focus on the finish line. I was addicted to the saltiness of green olives, the sweetness of Tang alongside a glass of whiskey in the evening.

. . .

On the Fourth of July we took the day off. I had reports to write, tires to patch, and oil to change in trucks and four-wheelers, but I left the ranch to go into town and meet Jenny and Carl at the bar in Wall.

Most young men and women in Wall were dispersers who didn't make it far from their ranch families. First sons get the ranch, stay on with Dad, and learn the ropes. Economics and tradition dictate that all others must move on. Those who couldn't yet muster the courage to go farther than the nearest town were now a mix of waitresses, busboys, cooks, seasonal construction workers, or day laborers at tourist trinket shops. They all drank hard; but Mandy's and Tim's blackened and rotting teeth showed they went a little further, dabbling with meth in the back alley behind the strip mall of shops on Main Street.

Courtney had become Jenny's new best friend because of two inevitable factors: she was the only other pretty single girl in town, and she hung out at the bar in the evenings. Most times men bought her the drinks, but this week she did the buying with money earned as a sign holder for the road construction going on just west of town. Today she brought along two boys from out of town whom she met on the road crew. Jenny was sitting at the end of the bar and was dressed as a different woman than the one I know from the basin. Her hair was sprayed wet to hold curls, her eyelashes were painted dark black, and tight jeans and a black tank top showed off her young curves.

"Wow, you really cleaned up. Who are you?" I said to Jenny.

She smiled and Courtney jumped in, "Don't worry, its not for you, its for Chad . . . later . . . he'll be at the bar in Quinn tonight."

Five years older than Jenny, Chad made a living by inheriting his father's small cow–calf operation east of town. He did a little better than normal by also starting his own trucking business, running steers and cast-off cows of neighboring ranches down to feedlots for fattening prior to slaughter. Jenny had met Chad six years earlier when he was a high school senior and she had spent a year living near Wall with her father after his divorce. Chad was her crush, and we could tell that Jenny, now a full-grown woman, meant to make up for lost time. She wanted to catch his eye after years of building on daydreams back home in Illinois.

The bar filled into the afternoon as we took on the sideways glances of tourists who avoided direct eye contact with our loud table of locals. By dusk we were thoroughly drunk on light beer and shots of tequila, some paid for but most given away. We made way for the paying patrons, and by dark we had shuffled back to the corner booth of the bar crammed in among the bright lights of slot machines on the back wall with bar staff on break chain-smoking cigarettes. There was no order to the chaos of conversation, stories, flirting, crap talked about people who were not there but should have been, or were not there and weren't welcome.

At some point, I don't know exactly when, Jenny leaned over to Courtney and gave the order to move our party to Quinn. Normally moving a drunk group of settled people from a bar is not easy. The noise, air, familiarity of a waitress, it all becomes accustomed and comforting as one otherwise loses one's senses in alcohol. Yet without any obvious scheming, in a way all bar girls know how to do, they put their plan into action. They made it playful at first, just suggesting the change of scenery, knowing no man wants to be told to leave, particularly a man drunk and comfortable late at night. But if a pretty woman suggests it, just offering the chance but acting like she has been talked out of it, and then waits ten minutes before forcing the action . . . the men will follow. Knowing all along that once they get past a drunk man's stubborn pride, the laws of nature will take over and wherever the girls move the boys are sure to follow.

After fifteen minutes Jenny got up and grabbed me by the arm. Before I could resist we were outside and I could only look back as the two construction boys who were too drunk to stand grabbed for Courtney around the end of the corner table. She squirmed free, then mocked them playfully as she headed outside to join us, car keys in hand. Carl stood up a little too fast and wobbled toward the door, followed by rodeo cowboy Nick, who was still dirty and worn out from the day's competition but swore he was sober enough to drive the ten miles out of town, so we packed into Courtney's worn-out four-door Buick.

Quinn was another quiet prairie town that had the benefit (or curse) of being just far enough off the interstate to lack traffic to a gas station as a sole source of income. The single unsigned bar on the only road in town was usually closed or vacant, but on the Fourth of July people from across the county packed into it and spilled out onto the dirt parking lot where livestock corral panels had been set up to outline the boundaries of a makeshift patio. We pulled off the highway and into the parking lot filled with orderly lines of pickup trucks. Inside, the bar was filled with handsome couples in cowboy hats and pressed shirts. Along the far back wall children were running circles around the old pool table from which the cue sticks and balls were long gone.

In the corner, Jenny had already found a group of men who piled around her. She smiled and loudly talked of saving prairie dogs as the men guffawed, halfway pretending to listen to her story of disease and Black Death. They softened as she told them of hours on four-wheelers with pesticide as they looked along her neckline and down her shirt, pretending to listen all the way to the end when she talked of hoping to see a ferret.

I found my old friend Doug at the bar. He gave me an appreciative nod, knowing that I had left the ranch house in my best outfit of a collared button-up Wrangler shirt and jeans. I had scraped my cowboy boots with a coarse scrub brush to at least be clean, because they were far past holding a shine. I left my worn-out straw hat behind, knowing it wouldn't complement the look but rather pretentiously suggest the look of someone who wanted folks to think he worked hard. I squeezed in next to Doug at the bar and we talked of gas and cattle prices, the weather, how Vanessa's bighorn sheep had left Badlands National Park after being flown up from New Mexico and turned up ten miles north at the seldom-used Wall airport landing strip. They had to be shooed off the runway so the ex-mayor could take out his Cessna.

In a hushed tone I changed the topic to prairie dogs. Doug told me that he had already heard the news about the Forest Service turning down the proposed management alternative of shrinking the Conata Basin's prairie dog population down to eight thousand acres. Instead, the agency opted for increasing the defined recovery area around the basin to seventy-nine thousand acres and expanding to another twenty-five-thousand-acre area. Ranchers were already threatening lawsuits, even though poisoning of prairie dogs continued on the boundaries as the Forest Service tried to be a good neighbor by keeping prairie dogs away from boundary fences.

At the other end of the bar, Carl was well into being drunk on hard liquor and had wedged himself into a corner seat next to an overweight man wearing an International Harvester ballcap. I didn't know him, but judging by jowls and nose shape his father was nearby. Carl leaned forward and voices started to escalate above the hum of the bar.

"Damn prairie dogs, can't get rid of 'em."

"What is wrong with prairie dogs . . . You damn ranchers are spending more money poisoning than the prairie dogs actually take from you." Carl paused only long enough to take another drink.

"Not that you have to pay for it, the government pays. Hell, the only way you stay afloat is with farm bill subsidies."

The scene went taut. It was obvious to everyone but Carl that he had broken the first rule of visiting a new town: Don't bring outsider ideas to a small family bar. Worse, Carl was particularly indelicate when drunk, and because of his obsessive weightlifting, always looked ready for a fight. Doug leaned toward me. "You better get your friend out of here before he gets his ass kicked."

I gathered Carl into the parking lot by lying to him that Jenny needed help, navigating past glaring eyes and back to the Buick outside. Courtney helped me push Carl in and then headed back to collect everyone else. As I held the car door shut on Carl, I saw Jenny in the corner of the corral talking with Chad. The two of them were close together, leaning on the temporary metal rails marking the far edge of the parking lot. Jenny looked down, kicking her boot at the ground. Chad stood tall next to her, looking straight into her brown hair, waiting for her eyes.

"She said she'll be fine," Courtney told me as she returned with Nick. "She'll catch a ride home later."

. . .

Back in Wall we found the construction boys and a few cowboys walking Main Street. The bars were closing, so we bought beer from the gas station and continued drinking in the parking lot of the Oasis Motel. The Oasis was a long-out-of-business, old-fashioned drive-up motel decorated with faded pastels on a corner lot. The pink cement water fountain out front was cracked and dry, now holding only brown, empty broken beer bottles. I sat on the curb while the others piled into a room for which Courtney had the key on nights like this when it was too late to go home. She lingered by the door while the two construction boys called for her to come inside.

"You can crash in the room with the boys," Courtney told me. She waited by the door looking at me. "You gonna wait up for her?"

"Yeah, I guess."

"Yeah, that Chad is bad news. He's just looking for a wife to have kids, sit at the ranch house, fit in and go to church."

"I know," I replied.

"She told me she would be back tonight."

The construction boys yelled for Courtney in unison and threw a shoe out the door.

"Shut up," she yelled as she turned and walked into the room. The boys bugled with excitement and bravado as she closed the door behind her. At the far end of the parking lot, Carl had started to stumble home but only made it to the nearest telephone pole where he stood, forehead against the pole, trying to steady his swimming head.

Finally alone under the clear night sky, I felt the alcohol slip away and I began to wonder what was left of lonely lives and wild places. The Old West was gone, now filled with dreams of satellite dishes and pickup trucks. Fenced-in cattle were hauled to feedlots in Nebraska, corn-fattened and even farther from the roots of this place. Beef tastes of the grasses you feed it, but today there is little of the mixed-grass prairie of South Dakota left in a steak once it is on the dinner plates of America. Layers of mythos had been laid down telling people this is how it had to be. How it was supposed to be. How you were supposed to live. What you were supposed to want.

I felt for first sons like Chad, inheriting the monarchy of running a dying ranch. Long days and single nights of trying to make ends meet. Hearing of high corn prices and contemplating plowing the best hay pastures. Contemplating selling out. Families fractured. Trying to find a wife, but working too hard on the land to play in nearby towns and cities where available young women were increasingly hard to come by. Seeing brothers and sisters go away to cities to get business jobs with health insurance and retirement plans. This was the modern West.

By 3:00 A.M. I was starting to become too sober to stay awake. I went into the hotel room with five single beds packed in. Inside, without street lights, the room was dark, filled with the comforting sound of deep sleeping breaths of a half-dozen bodies, both fully and half dressed occupying the four beds to the back of the room. I lay down with my boots on in the only bed unoccupied, a small naked single-bed mattress closest to the door.

I put my head down and found the clarity of thought that comes just before sleep.

> How am I to justify my solitude? I am a generation too late to work among Schaller, Murie, and Douglas-Hamilton. The iconic age of naturalists working, thinking, and documenting wild things in wild places is gone. There are few places far from a road, no place beyond human touch. What is to define my age in wildlife biology? I hope this is the age of restoration. Restoring the character of the land, what distinguishes it apart from the rest of the increasingly globalized and homogenized world. Just west of the Conata Basin Dan O'Brien and other ranchers are switching from cattle to bison, trying to be on the forefront of agricultural markets by creating interest and producing products that are desired, rather than by reacting to deeply rooted family beliefs. They are attempting to shift dinner table daydreams from selling out or plowing up the prairie, to bison, elk, prairie dogs, and black-footed ferrets.

Headlights came off the highway and into the Oasis parking lot. A shiny white Jaguar with a silver cat leaping from the hood pulled up to the curb. Jenny came out of the back saying good-byes and walked into the room, stopping just inside the door. She surveyed the room, letting her eyes adjust to the darkness, trying to use the street lights to choose a bed for the last few hours before morning. I pretended to sleep as she closed the door behind her and sat on the end of my bed, taking off her shoes.

"Nice car," I whispered.

"Its Chad's sister's, she's up from Denver for the celebration."

"Well, you made it out safe. Are you still in love?"

"Love?"

"Yeah, love. Are you a couple now?"

"I guess." She paused at the end of the bed, mulling over the word *couple* or *love* or whether she chose the right bed, surprisingly sober.

"He wants me to slow down, go to church with him Sunday. . . . He wants to marry me."

She lay down beside me, on her side, facing away.

"Marry you. He said that already?"

"Yep."

"You're not ready to be married. Are you?"

I felt her exhale and close her eyes.

• • •

On the last day of work for the season, I sat in the pickup truck trying to delay the coming end. Clay-colored swift fox pups played by the brightly colored orange flagging that marked the extent of the prairie dog burrows we had dusted and where we planned to stop, in this far northwest corner of the basin. Watching a family of five nipping at each others' ears and wrestling at the den entrance, I felt like I was witnessing another biologist's lifetime achievement. These were first-generation South Dakota foxes from parents translocated by pickup out of Kansas, Colorado, or Wyoming and released by Doug to this centerpiece of the Great Plains. They were succeeding here, but although our little group of three burrow dusters visited more than eighty thousand burrows over the span of five months, we were protecting only fifteen hundred of the thirty-one thousand acres of prairie dogs in the basin. It was barely enough for four female foxes to raise their families. I wondered what they would eat when plague comes through and wipes out the remaining prairie dogs. How far would they have to disperse before they found a new home?

I already knew that I had to stop studying the ecology and biology of ferrets and start studying their diseases. Emergency and disaster had led me to give up the romantic notion of being an ecologist who spent a lifetime studying the behavior and function of rare carnivores in the wild. There was no more time for that. Rather than learning from them during long nights of observation, I had to simply respond to their needs, the needs of a few hundred ferrets in Conata Basin, a few of which I had come to know well after three years of intensive study. I had to refocus my life on saving the leading example of success in the ferret program. We were combating the disease without fully understanding it, following a path that started in the trenches, using the only tool we knew of to control the spread—a flea powder that was applied on a scale not by the pet store bottle, but by the pallet load directly from the pharmaceutical manufacturer. We were committed to a cycle of dusting that needed to continue in perpetuity for there to be hope of keeping plague at bay, and ferrets on the ground, year after year.

I stepped out of the truck just as Carl turned off his cell phone. He had called the head office in Pierre to check in and tell them of our progress.

"Plague was confirmed just ten miles south of us down on the Pine Ridge," Carl said.

"Really?" I said as I started loading gear for the day on my four-wheeler. "Guess it won't be long now."

Jenny didn't stop working. She ratcheted down straps holding the equipment onto the luggage racks of her four-wheeler, checked the gas, and refilled the tires because cactus spines had made small holes that leaked air during the night. We were all silent, keeping to our morning work ritual. When Jenny finished her preparations, she stopped and looked up at me and Carl. "Chad says the plague is natural . . . and we should just let it run its course."

"You know plague is from Asia, introduced by us, about as natural as Velveeta cheese," I replied.

We drove out onto the prairie, picking up where we left off on the previous day. Without speaking, we started treating each burrow with five grams of powder. After months of working together we had developed a system of bobbing and weaving that had been honed for maximum efficiency as we moved forward across the prairie. Always forward. How could she ignore all of this? We fail when the worst parts of human history are allowed to repeat themselves. Was it wrong to think that persecution and plague were just natural conflicts in the inevitable and needed decline of prairie dogs, like the persecution and diseases that removed Native American Indian communities from these same plains 160 years ago? Influenza, syphilis, and smallpox were exotic diseases brought over from Europe and Asia just like the more recent invasion of plague. They all operated on the same principle of infecting new world residents that had not coevolved with the disease long enough to develop even a rudimentary level of immunity.

We knew that the plague bacteria were fighting for persistence like any higher-order organism, needing hosts that were both susceptible and abundant. We also knew that these were two attributes for which the Conata Basin and all other ferret reintroduction sites were ideally qualified. The Conata Basin in particular was attractive because of its large and dense prairie dog population. The very traits that made it so successful for ferret recovery were exactly what made it so vulnerable and appealing to plague: the horrible irony of how a rare, sensitive native found the same benefit as the exotic, seemingly unstoppable bacteria. By dusting, we thought we were keeping the disease at bay, fighting the seemingly noble fight to protect the natives, preserve the past. But how long could we keep this up?

While our team of three worked in Conata Basin, the scale of burrow dusting across the Great Plains was immense. Every year thousands of acres of prairie dogs were treated across at least seven western states where there were ongoing efforts to conserve them. Most efforts focused on ferret reintroduction sites or prairie dog colonies located where there was concern over disease transfer to humans or house pets. The efforts were focused on keeping flea populations below some critical threshold needed to provide enough infected bites to cause an epizootic. More than half a million burrows were treated each year using thousands of pounds of flea powder.

Ferrets were thought to be resistant to the disease like polecats of the Mongolian steppe, but in actuality they are highly susceptible, perhaps more so than prairie dogs. In a series of studies in Montana, Randy Matchett and Dean Biggins found that ferrets declined even before epizootic die-offs in prairie dogs occurred. So keeping the blood-sucking vectors below this threshold might not be enough for sensitive ferrets, making control of plague even more of a mystery than simple proactive flea control.

Travis, who had been monitoring and studying ferrets in Conata Basin since they were first released in 1995, knew what plague could do to the ferrets in the basin and had taken on the challenge of plague in the best way he could, by saving the ferrets directly. He got his hands on a trial vaccine for black-footed ferrets derived from the one given to troops in the U.S. Army. Because it had to be injected, Travis needed to capture each ferret in the basin at least once a year, and preferably twice a year to give a booster. The annual ferret roundup he orchestrated took on a whole new level of intensity. Yet there was no vaccine for prairie dogs, and no practical way to inject every prairie dog or even enough prairie dogs to save ferrets from extirpation from lack of food. So we all knew that if plague hit the basin, and the prairie dog population crashed, inoculation of ferrets would not matter if the prairie dogs on which they depended did not also survive.

Even with the reality of a plague epizootic waiting on the doorstep, Travis persevered. Ahead of the plague, there was value in vaccinating ferrets, for as Randy showed at UL Bend, even before the full epizootic hits, some lower level of plague could be picked up by ferrets and decimate their population. So, each year, Travis continued trapping as many ferrets as possible starting in late summer and running into late fall, vaccinating hundreds of animals in a last-ditch effort to save his life's work.

. . .

For the past few weeks Jenny no longer stopped to watch a rattlesnake consume a prairie dog or to count the eggs in the nest of a horned lark. She started wearing headphones to listen to music while working, and I no longer caught her lingering to observe a ferruginous hawk in flight. She had signed on to be a waitress in town after this job ended, and she moved with the clear conscience of an hourly employee. Within two years she would be pregnant and married, a permanent resident of Wall.

"You still want to see a ferret?" I asked.

She was silent.

"You should come out with me tonight after sundown."

"Its Harley's birthday tonight," she replied, as if I should have remembered.

"Who is Harley?"

"Harley," she enunciated with a smile, "the bartender."

"Do you really need to be there?" I asked.

"Well no, but it will be fun. Can't we do it Thursday night?"

"Nope. Timing is perfect tonight. It's now or never."

After sundown we drove out of town, down the paved zigzag road over the Badlands wall, onto the familiar dirt roads and through fence gates leading into the basin. Once in the pasture where we had spent the last three months memorizing every burrow and badger hole, we drove around in expanding circles, scanning the prairie with a powerful Australia-made spotlight developed for kangaroo hunting. The moon was not yet up and there was only the faint glow of lights in Rapid City, fifty miles to the west. I turned on the radio and it told us of baby tigers confiscated from carry-on luggage at the Bangkok airport. They were drugged and stored in a suitcase next to stuffed toy tigers, supposedly to "hide the animals" from authorities before transport to a buyer somewhere overseas.

How simple our conservation problem is, I thought to myself. We had already figured out that the ferrets needed one thing: more prairie dogs. They needed populations even larger than those that existed twenty years ago, prior to plague coming into the plains, populations like this one in the Conata Basin.

That night we didn't know that next spring, on May 13, 2008, disaster would strike the ferret recovery program. Plague would make the jump up to the basin, starting the process of killing every prairie dog not treated with flea powder. It would mean that the very population that was being hemmed in on its sides by government-sponsored poisoning campaigns would soon be rotting at its core down to a fraction of its former self.

We didn't know that we would both be back next year working with an urgency even further on verge of exhaustion. We didn't know that the contractors for hire who poisoned prairie dogs with toxin-coated oats in the fall would be hired in subsequent years to do our dirty monotonous job of burrow dusting, shifting their schedules so they were in the Conata Basin for much of the year, protecting prairie dogs on one side of a fence in the summer, then killing prairie dogs on the other side of the fence in the fall. This was ideal income security, keeping some prairie dogs alive and keeping them rare enough to conserve.

After only two hours and three loops around the pasture I locked onto the green eyes of a young female ferret in the spotlight beam, and Jenny was entranced.

"Is that one?" she asked.

"Yes."

Flashbacks from her college mammalogy class told her that this was one of only a handful of places you could see such a critter in the world. For a moment I saw that young girl before she was exposed to bars and men in Wall, Interior, Scenic, Quinn, and points farther south on the reservation.

"Do you want to get closer?" I asked.

"Can we?"

I slowly moved the truck closer to the animal and she shifted to my side of the bench seat, leaning over me to look out the driver's side window and at the ferret by the burrow only fifty feet away. The smell of her shampoo mixed with prairie smells of dust and crushed fetid marigold wafted into the window from our sudden stop. It was just the day before that she told me she had decided to stay in Wall for the winter and find a job.

"You should move on with me," I told her.

"Don't get your hopes up," she replied without looking up at me.

I looked out the window with her at the ferret peeking from the prairie dog burrow as I had seen ferrets do hundreds of times. Each one slightly different in personality, some tolerating my presence, others bobbing up and down with concern. I wanted to tell her that she had grasped onto the wrong part of the West, that there were no Old West American cowboys left here. That European civilization had changed this area for the worse, damaging nature and human community to the point of collapse. That the place she was now becoming a part of needed to change. That without change this place would become biologically and socially indistinguishable from the barren corn fields in Illinois where she grew up. And finally, that we needed smart, young, educated women like her to make change happen.

"You don't have to marry me, just get out of this town."

She leaned back to her side of the truck and I looked at her in the face for the first time. She was silent, looking forward through the windshield, ready to head home. I suddenly realized that it was too late; she was gone.

"You should go on, do great things."

"Who says I want to do great things?"

CHAPTER 12

Kansas

Mike Lockhart and I were finishing a tour of Texas, stopping in Marathon before heading up through the panhandle to visit some ranchers Mike knew through someone, somehow, who said they might have some prairie dogs left on their land. The next week I was down in Arizona, visiting a large ranch just south of the rim of the Grand Canyon. We were continuing our tour of states where plague hadn't completely decimated (or had at least left remnant) prairie dog populations—hunting for even small pockets of prairie dogs in remote corners of Utah, Colorado, and New Mexico that might be suitable for a few ferrets. Fighting for the minimum.

Early in May, I traveled with Mike to Manhattan, Kansas, for a meeting about getting ferrets into that state. As we drove due east, the green and humidity reminded us we were no longer in the Mountain Time Zone. The arid mixed-grass prairie was giving way to rolling tallgrass prairie that was still partly green in midsummer. I spent the night at Sam Wisely's place north of town, drinking tequila and catching up on her new professor status. We reminisced about Chihuahua, UL Bend, and other ferret sites as field biologists are likely to do when the field was no longer where they played, lamenting how we were now increasingly strapped to computers and reports, and in her case, students.

The next day, we met at the local U.S. Fish and Wildlife Service state office to talk about potential ferret sites in Kansas. Because most of Kansas is locked up as private land, unlike the larger federal land holdings out west, there were no suitable sites on publicly managed lands. Ted Turner, the media mogul turned conservationist who was one of the largest private landowners in the United States, owned a ranch in the eastern part of the state where he had restored bison and prairie dogs. Yet unlike his ranch in New Mexico that already had ferrets on the ground, and another along the Bad River in South Dakota that was in line to begin receiving ferrets in a few years, his Kansas ranch had a prairie dog population that was still too small. They even had to mow the prairie to help prairie dogs keep up with the fast growth of the Midwest tallgrass prairie. The Nature Conservancy owned a ranch in the western part of the state, but was afraid of attracting further attention to their small prairie dog population because the manager already was dealing with angry neighbors.

Despite hitting the bottom of our usual list of federal and conservation organization partners we found hope offered by the owners of two private ranches near Russell Springs in western Kansas who were ready to buck the system. Gordon Barnhardt and Larry Haverfield had told the Fish and Wildlife Service they still had a couple of moderately sized prairie dog populations on their property and were open to releasing ferrets. This bold move resulted in the local Logan County Commission passing a resolution that forbade the reintroduction of ferrets. Because the resolution had no legal force, the commission followed up by resurrecting a homestead-era state statute that prohibited private landowners from allowing prairie dogs on their land. The county would deal with violators by hiring poisoners to go on the property and exterminate the prairie dogs, then hand the bill to the landowner.

Logan County had a deep past with prairie dogs and figured prominently in C. Hart Merriam's influential 1902 article "The Prairie Dog of the Great Plains." The original short, eloquent report provided details of the ecology of prairie dogs and outlined a government policy for their eradication. The evidence for Merriam's case was largely based on stories from Logan County, where he reported that the establishment of farms and cattle ranches in the area by "white settlers" during the 1800s had resulted in an increase in prairie dogs to the point that the whole southern half of Logan County (estimated then to be around three hundred square miles) was "one contiguous dog town." Merriam offered this in his manifesto as an instance when prairie dogs had "overrun the range," causing families to abandon their ranches and leave homesteads vacant. He poetically highlighted this varmint-caused human exodus by linking it to the closing of a post office in Elkader, Kansas.

His story would even persuade East Coast politicians to support a government-sponsored poisoning program to eradicate prairie dogs that would last over a century and continues to this day.

Mike had already arranged a meeting with Larry Haverfield, and after finishing in Manhattan we drove back west, nearly to the Colorado border, and spent the night in Oakley, Kansas. The next morning, we drove out to Larry's ranch, where he greeted us in his driveway. His dirty ballcap, long grey beard, and denim overalls gave him an inherent wholesomeness even before he opened his mouth. We stepped out of the government truck and Mike gave him one of the framed prints of a polar bear photograph that he took during his last trip to Alaska to help with polar bear surveys. This annual spring trip gave him a few weeks away from the demands of the ferret recovery program, and he photographed beautiful white-on-white images as he helped track, dart, and monitor the status of the large carnivores while helping a friend during his helicopter surveys. I had seen him give a similar framed print to a biologist in Texas the week before, and I interpreted them as offerings to their recipients for taking the time to meet with us, and also for the thankless sacrifice anyone becoming involved with ferret recovery would soon have to make—putting a conservation ethic that involved prairie dog conservation above persecution by one's peers and the local public. Larry took the framed photograph inside, put it on his bookshelf, and offered to give us a tour.

Driving around his property through the maze of gates and fences in the rolling Kansas hills, we saw his place swarming with ferruginous hawks and other raptors. We were used to talking to prairie dog activists, typically from Denver or other big cities, who became committed to conserving prairie dogs the same way as others become concerned for human rights. Prairie dog huggers, some would call them. But Larry was not just into prairie dogs for prairie dog sake, he was enthralled with the whole system. And contrary to prairie dog huggers, Larry would shoot prairie dogs from his front porch with a .22 rifle. He would take only one or two, and then get on his old motorcycle and drive to the nearest swift fox den on his ranch to feed the young kits.

"I can still shoot prairie dogs if we put ferrets out here, right?" Larry asked as we drove down into a valley on his property. "So I can feed my swift fox friends."

"Absolutely," Mike said with a smile.

"They tame right down and they have had litters in the same burrow the last three summers."

Mike explained to him that ferrets den in burrows to raise kits in the same way.

"I guess I could feed the ferrets too."

Mike and I smiled, knowing there must be some federal law against it, but we didn't say anything. We knew Larry had managed prairie dog populations on his property for decades in the face of extreme social pressure to exterminate them as his neighbors had all accomplished decades before. He would never shoot so many as to harm the population and limit ferrets from taking hold.

We drove to the pasture where, last summer, he strategically moved his cows to keep the county from poisoning on his land. The presence of the cows was the only thing left to stop the county from putting out poison oat bait. They halted under fear of the backlash that would occur as a result of a hundred cows that would be sick or dead from eating the poisoned prairie dog bait. The county asked him to move the cows so the professionals they had hired from Wyoming could go in and kill the prairie dogs. He told them to shove it.

There was still a pending lawsuit that made Larry cringe a little when we talked. Like any western rancher, he didn't like bringing in the lawyers, but he liked the idea of giving in to the county officials even less. In truth, he wanted to make it work with his neighbors. He drove us to the boundary of his property and showed us the ninety-foot buffer strip that he kept free of prairie dogs by putting out a narrow belt of poison every year. He'd also erected an electrified boundary fence to keep cattle away from the buffer strip, allowing grasses to grow high and create a natural visual barrier that would dissuade prairie dogs from trying to leave his land. I was amazed; while running a ranch he was also doing what took crews of federal biologists and contractors to accomplish. Efficiency.

We drove back to his house set against the hillside in the middle of his property. His front porch railing was covered with skulls of coyotes, prairie dogs, and other small animals. He invited us in for lunch, and we stepped through the doorway and into his simple home. It had wood paneling and a low-hanging ceiling that kept the inside dim even in the midday sun. Old linoleum tile in the kitchen led to a small living room with two recliners facing an old television. Beside the television and pictures of children and grandchildren were bookshelves containing textbooks on cattle ranching. Apologizing for his wife being out of town, Larry started up the stove and put a steak he had defrosted for us in the pan. He saw me eyeing the books and walked over to pull one out.

"I went to a workshop back in 1986 that changed my life and how I raise cattle."

His eyes brightened and voice quickened as he expounded on the benefits of rotational grazing to the health of the grasslands and the weight gain of cattle. He actively followed the practice by putting up fences to make a checkerboard of pastures where he could keep tight control of where cattle grazed. By forcing cattle to intensively graze a given area to a low stubble, he said, you could gain the benefits of trimming grasses and increasing nutrient content similar to the seasonal grazing systems where cattle are allowed to roam freely across a larger area. But the advantage of a rotational grazing system was that you could exclude cattle after moving them to a new pasture, resting the pasture and allowing the grasses to recover. Despite my aversion to fences, within two minutes I was convinced that if I ever settled on a patch of ground, I would split the land into small, fenced pastures and follow his model.

Taking the steak out of the cast-iron pan, he cut the oversized slab of meat into thirds with a fork and knife and slid pieces onto dinner plates.

"That's 100 percent grass-fed beef," he said. "You can eat the fat and all. It's leaner, richer, than the feedlot cattle you guys are used to."

We nodded in agreement. It wasn't the thick, tender, fatty slab of meat from a corn-fed feedlot steer or some beer-fed, Kobe beef filet. This steak was seasoned with the native grasses Larry had nurtured and allowed his livestock to ruminate on in his still largely native and unplowed section of the Great Plains.

"Never had a better steak," Mike said, and I agreed.

• • •

In 2008 I returned to the Conata Basin after the spring rains to start up the burrow dusting season. We were already in full disaster mode, with plague taking out hundred-acre chunks of prairie dog colonies each week. We knew we could not stay ahead of its march north and west. Our only hope was to make a boundary of dusted burrows akin to a fire break. If the boundary held and stopped the spread north long enough, we could spend the remaining months dousing the remaining burrows behind the line.

To do this we once again summoned all hands on deck. Biologists, firefighters, and college students on summer break were hired, trained, and thrown on the back of four-wheelers with dusting machines. Dousing burrows in a 150-foot-wide swath on the open plains of the basin, we worked our way west, and on reaching the other side, turned to work back east. We made a dozen parallel lines each day as we moved south to north. The goal was to eventually hit every burrow in the thousand-acre section we hoped to protect. Progress was painfully slow and the pace was grueling, no days off, working dawn to dusk and staying up into the night to fix equipment broken during the day: slender aluminum gears stripped smooth from use in our fancy, custom-made thousand-dollar dusting machines; four-wheelers that had gone through tires that leaked air continually from running over prickly pear cactus spines.

Two weeks after starting, Carl had quit because there was just too much to do and not enough people. After two weeks most people grew tired, quit, and moved on. They wilted in the July summer heat and the prospect of seemingly endless hours on the back of four-wheelers in the cloudless sky. Their skin baked to a dark brown. Sunglasses, bandanas, and masks kept the prairie dust and flea powder from choking our lungs and eyes as we wove around each other on the dry earth. A biologist from Colorado stayed an entire month in the basin with us, and then in the middle of one day had an epileptic seizure. The heat and lack of sleep brought on an episode for the first time in more than ten years. He was evacuated by ambulance, then we continued on.

New waves of volunteers and employees were ushered in as others were worn out or asked to go back to their normal jobs. I plugged on, trying to manage the personalities that increasingly dallied in between mile-long transects. By early July, afternoon temperatures flirted with 100 degrees, and breaks for water turned into half-hour conversations as everyone tired. Talk was of beer and bars that we knew we wouldn't have energy for, and of what we would do when we finished the summer. There was never talk of the next day, never talk of the plague. That was for others to monitor, study, and worry about.

By the end of July we were nearly finished with the first major patch of prairie dogs on the basin's eastern edge. For the first time, we could see a sign of accomplishment before we started the next pasture the following day. When temperatures were predicted to get over 100 degrees in the basin we tried to start early, driving out before dawn. But today it was 100 before noon. Too hot, we finished our canteens of water for the day by 2:00 P.M., but we saw the finish line of the basin perimeter fence line only a couple hundred feet away and pushed through, skipping breaks in order to finish and get out of the sun. When we reached the end we started up our trucks and hurried to stow our gear, anxious to jump into the cabs where air conditioners were running at full blast.

We drove up and over the Badlands wall, through the park and back to Wall. We were all badly dehydrated. Our cooked brains had no thoughts, no conversation. We were numb to the world. I dropped everyone off at the Forest Service office, and we dispersed to our hotel rooms for the night. Needing water, I went to the local gas station that doubled as a sandwich shop. I stepped up to the counter and ordered a sandwich, any sandwich, I didn't care. I also ordered an extra large drink, the ridiculously oversized American gas station fountain cup that required two hands to fit all the way around. I filled up the cup at the soda fountain with iced tea, drank it down, filled it again, and drank it down. My sandwich was ready so I paid for it, scrawled my name on the credit card receipt. I could focus only on the straight blank line at the bottom. All else was a blur. I'd pay any amount. It didn't matter. My mind began to spin from too much of the sugary water too fast. I needed to sit down and steady my head before driving again. I needed to close my eyes to stop a spinning world. I sat in a restaurant booth, lay my head down, and instantly fell into a deep sleep. I woke to a feeling of relief, only slowly realizing that I was urinating in the seat. I raised my head and regained consciousness, finally being able to stop the flow. I checked my crotch and then the floor, and found it was only enough to saturate my jeans, cool the skin on the inside of my thighs, and moisten the top of the plastic-covered chair. I didn't know if I had been out for five seconds, a minute, or an hour, but for at least a few seconds I felt nothing, and nothing felt good. Now I was awake and exhaustion mixed with embarrassment. I didn't want the sandwich; I just wanted to drink more but didn't want to risk a repeat performance, so I shuffled out without filling my cup. Back into the dry heat and home to my rented room to rest before another day.

. . .

In early August, I skipped out on burrow dusting for a few days and headed to Wyoming to rendezvous with Mike. We were to meet with the Northern Cheyenne Tribe just across the border in Montana. They said they had a few hundred acres of prairie dogs near the town of Lame Deer that were growing back after a plague epizootic. If they were to get ferrets on their tribal ground there was the potential for federal grant money to come their way to conserve prairie dogs, to hire people to dust against plague and spotlight for ferrets, and to buy off some ranchers with incentives. Mike said it was worth a look.

Driving west from South Dakota, I saw the cool forested peaks of the Bighorn Mountains. I was to meet Mike the next morning for breakfast in Sheridan, but after months of hotel living I did not want to stay in another modular room. I was tired of the hot days on the plains and of sleeping in the cocoon of an air-conditioned hotel room at night. Coming into Sheridan, I turned off the interstate, filled up on gas, and skirted downtown by driving south and east toward the base of the mountains. I found a gravel Forest Service road with a sign that advised four-wheel-drive vehicles only, and started the climb up. After weeks on the prairie, it felt good to head toward the cool of the mountains, away from my planned travel route. I entered a land of weekend campers and travel trailers set up for four-wheeler excursions on the maze of dirt roads cut by loggers and the government over the generations and then left to reclaim themselves. I wanted to find a shaded cool spot, become a hermit in the hills.

I drove up narrower and narrower roads that became little more than rubbled paths as I headed uphill. Shifting into four wheel drive, I continued uphill to high meadows with pools of water. I heard the roar of small engines coming downhill and passed two four-wheelers returning down to their camp in the valley below. I went farther, finding a small lake and parking on the far side in a stand of pine trees. In my rush, I had forgotten dinner. Clouds rolled in with a mist. The weather turned cool and large raindrops fell, slowly at first, and then in a storm of rain that washed the South Dakota dust from the truck. Digging into my backpack behind the seat, I found a granola bar and a bottle of water. I reclined the seat, ate, and waited out the storm.

After a half-hour the heavy rain passed and I rolled down the window. Humid air filled my lungs. It was 8:00 P.M., but still light out, too early to sleep. I heard a clap of thunder to the east. The storm had passed. Outside, a bell rang in the mist out of view. A cow bell, then another, moved toward me. Then the source appeared out of the cloud in a soft afternoon light, a dozen horses grazing slowly forward in the wet grass below the lake where the water overflowed.

The cool air and moisture reminded me of the color green, and of other places I had lived and known. Two years of burrow dusting in the Conata Basin had drained me. This life on the road was no good, fighting a losing battle, driving myself to incontinence and near insanity year after year. I needed to recharge and regain hope.

. . .

By January I was in Salt Lake City at a meeting of managers from across the West to discuss prairie dogs, plague, and ways to combat the disease. There was talk of a plague vaccine for prairie dogs that could be administered in bait form in peanut butter–flavored tablets. Land managers

FIGURE 22. A team in central Montana applying insecticide to prairie dog burrows to kill fleas that could transmit plague. Each year, more than half a million prairie dog burrows are treated with insecticide across the Great Plains.

hoped that it could one day be mass-produced and dropped from airplanes to inoculate acres of prairie dogs. But the vaccine would take time, likely a decade or more before it was on the ground. There were fears that it would be too late or too costly to be done in any effective way. We needed something now, something easier than dusting burrows.

Once again we turned for help from the profession that has made an art of killing prairie dogs. A small company in Colorado was rumored to make a rodent-killing bait with an insecticide mixed in, so that moments before the rodent itself died, the fleas and ticks were killed. Their clients typically were city managers in California who wanted to prevent apartment buildings filled with rats from becoming infested with hungry fleas in search of new hosts. A small change in the recipe to remove the rodent-killing chemical could result in a bait that, when consumed by rodents, flowed into their blood stream but only killed the fleas. If it worked, an entire population or colony of prairie dogs could be cured of fleas not by dusting every burrow, but by scattering treated grain bait across the prairie for prairie dogs to consume on their own.

I saw my opportunity. After years of burrow dusting I thirsted for the organized approach and certainty of science, to head back into the field, conduct research, and provide an answer. I proposed to study whether this internal or systemic approach to flea control would work and stop plague outbreaks. The managers in Salt Lake City sounded enthusiastic and offered some money if we would do the work in southwestern Utah. I took on the project and went farther west into unfamiliar red rock country, Ed Abbey's country. A place on the westernmost edge of the range of prairie dogs, where they were likely to have been exposed to plague the longest as it migrated eastward from San Francisco. These prairie dogs were also the rarest, the federally threatened and locally named Utah prairie dog.

Utah prairie dogs are survivors. A large proportion of them live in the Cedar Valley on the edges of human society, surviving in the green irrigated gaps of agricultural fields and on the edges of the golf course just west of town. They thrive on these places, but at the cost of being persecuted or occasionally forcibly relocated farther east into the high desert of the Colorado Plateau, where there are vast public lands and fewer neighbors to complain.

It is this strategy of moving Utah prairie dogs onto government land of the high desert that was expected to recover the species. Yet when prairie dogs are relocated, less than 10 percent typically survive to the next year, so it became necessary to move a large number of individuals to create or restore a colony. The problem was that these high-elevation colonies were more prone to plague, the very factor that decimated them in the first place, further reducing the chance of establishing Utah prairie dog populations in the high desert. If we could deal with plague, then we could potentially restore Utah prairie dogs on public land and take the pressure off private and city lands down in the valley.

By May, I was in Utah trapping prairie dogs. Apart from the valley populations that resided at around fifty-five hundred feet, it was difficult to find even two populations of more than fifty individuals to study in the high desert. These populations that lived at around eight thousand feet were active for short periods of the year, and typically occurred at very low densities and in small colonies that were spread out across several hundred square miles. Most colonies contained only a dozen or so prairie dogs, some upward of fifty or a hundred. I wondered whether there were enough of them for me to study.

With the help of Nathan Brown, the State of Utah's Utah prairie dog expert, we settled on two prairie dog colonies; one just outside of Bryce Canyon National Park and another fifty miles north on the Awapa Plateau. The colonies were in arid grasslands atop benches overlooking

FIGURE 23. Researchers collect fleas from a captured Utah prairie dog *(Cynomys parvidens)* prior to releasing it back into the wild.

red- and orange-colored canyons. Nathan hired a crew of four prairie dog trappers and we camped in the shade of pines on the forested edges near draws and ridges, packed in water and food, and slept soundly under the desert night sky. Heading to the colony at dawn, we set traps and checked them at hourly intervals thereafter, hoping to capture the wary animals. When they were caught, we used anesthesia while we ran flea combs through their fur to see whether the systemic product was working. With tweezers, we collected the dislodged fleas in a small plastic tub held underneath the sleeping prairie dog, tallying the number of fleas on each prairie dog from treated and untreated sites. Like the scorekeeper at a basketball game, as the days unfolded I tried to do a tally in my head of how the whole experiment was going. We were still findings fleas on the treated prairie dogs, but it would take months of data analysis to detect whether there were mathematical differences between sites. Perhaps the bait didn't completely work, but did it kill enough fleas to limit the transmission of plague? We knew that different species of flea were better or worse at transmitting plague. What species were there in our small vials? Did any of them contain the plague-causing bacteria? It would take two years of fieldwork and hundreds of hours in the lab looking through microscopes to find out.

. . .

Two years passed in our study on Utah prairie dog plague ecology, treating prairie dogs and sampling their fleas. We stored and shipped thousands of fleas from hundreds of prairie dogs back to our laboratory for identification and testing. I was forced from being a carnivore ecologist into being a novice parasitologist and epidemiologist, developing theories on plague ecology, how and when plague outbreaks occur, what flea species were likely at play, and how and why the high-elevation populations up on the Colorado plateau were more susceptible to plague. These were educated guesses that, in the end, would need to be boiled down and compressed for the ears of managers.

With data analyzed and processed into charts and figures, and with thoughts and ideas compressed into hypotheses and theories, I wrote up my results to be published in an obscure scientific journal. The systemic grain bait worked at controlling fleas, but only intermittently. It worked better at some sites than others and on certain flea species and not others. Three sites were lost to plague during the two-year study, showing the bottom line that our experiment with the systemic cure was ineffective. We were still losing prairie dogs and needed to act fast.

I drove down to Cedar City to present my findings to the seasonal meeting of managers, state and federal biologists, and Cedar City commissioners who were all concerned about how to accomplish the federally mandated task of recovering the endangered Utah prairie dog. I told them that Utah prairie dogs were in imminent danger; that despite controlling fleas, we lost half of our high-elevation sites during our two-year study of plague. We found that flea control was likely not enough, and that there were more subtle factors at play. Flea species diversity was much higher than expected, complicating efforts to understand their interplay with plague and prairie dogs in the diverse grasslands of the high plateau. We suspected that other small rodents that were more resistant to plague might bring their infected fleas into the prairie dog colonies. In short, that there was much more to learn to be able to fight this disease effectively.

When I finished and opened the floor for questions, a few of the meeting participants were concerned, but most were dismissive of my warnings. Like the many ranchers I talked to in South Dakota, they thought that plague was a natural phenomenon helping their cause. Here in particular, after twenty years of exposure to reports on the disease, they were used to stories of high-elevation prairie dogs dying. Perhaps deep religious beliefs—that God created all things and that this disease was one of them—helped them reach this conclusion. They likely believed that plague had been put here now to do this to the prairie dogs, and it was inevitable regardless of our efforts. Others simply had by now accepted that the prairie dogs would continue to die in the high elevations until a better solution was found by science in a laboratory, not on the ground. I was still unable to take such a long, detached view of the disease, and I wanted to take action now, but there was more concern at the meeting about the next item on the agenda: a story in the local paper about a handful of prairie dogs moving into the Parowan town graveyard. I saw county commissioners counting votes in their heads as they imagined readers of Sunday morning papers cringing as they ate their breakfast and drank decaffeinated coffee. They quickly shifted their attention to calling for the eradication of rodents that were burrowing into the sealed caskets, a sure-fire political platform for reelection. Other committee members still dwelled on the perennial problem of how to trap more prairie dogs off the golf course on the edge of town.

After all the time and effort put into studying plague and fleas and prairie dogs, I felt the failure of an academic. Scientists tend to focus on puzzles in the natural world, fixing biological problems with biological solutions. But even with solutions, human tolerance would likely not have changed even if plague were out of the picture. There would always be a perennial fight with prairie dogs. I drove out of town and up to one of our old field camps outside of Bryce Canyon National Park. The prairie dogs had died out here over a year ago. We trapped and released them one day, and then the next they were gone or lying on the surface of the ground, some of them still in the last gasps of death, paws curled and eyes pointed up to the sky.

I thought of the ferrets that lived far to the east on much larger colonies of prairie dogs where the cycles of plague were just starting to take hold. Epizootic die-offs were followed by slow recovery over five or maybe ten years, and then more epizootic die-offs, following a law of diminishing returns until only small pockets of prairie dogs were left on a broken landscape.

It was hard not to see that a plague vaccine for prairie dogs might be created in our lifetime, and that by vaccinating them each year it could work to create small, high-elevation pockets of Utah prairie dogs to recover the species. But what about ferrets, which required 10,000 acres of prairie dogs? And not just one 10,000-acre site, but ten 10,000-acre sites if the species were to be removed from the endangered list?

With more than 262 million acres of publicly owned rangeland in the United States, ferrets would require only 0.03 percent of it to contain prairie dogs. That is 1/3000 of western publicly owned land, but in Colorado, Montana, Utah, South Dakota, and elsewhere I still encountered federal wildlife biologists who believed that prairie dogs were varmints to be controlled and exterminated. Human attitudes can change, over decades or generations, as shown by views on slavery, women's suffrage, gay rights, and animal rights. But the challenge with prairie dogs was that they needed not only to be tolerated, but needed to be conserved proactively with vaccines or flea control or shifting grazing practices so that they survived and flourished.

Perhaps just as in Utah, the future for prairie dogs lies in intensively managing small populations using practices that verge on farming them like cattle or bison, with vaccines and grassland management regimes that maximize prairie dog density and abundance within defined areas. Perhaps the future for ferrets does not lie in the vast public lands and uninhabited corners of the Great Plains, but in the active hands of private landowners like Larry Haverfield in Kansas.

I cringed as I thought these domesticated thoughts in Ed Abbey country, imagining him rolling over in his shallow, unmarked desert grave somewhere to the southeast in this red rock desert. Taking such a path would harm the classic image of ferrets as rare wildlife on wide-open public lands in the western United States. But perhaps saving ferrets would require us to rethink the value of wild away from the Edward Abbey and John Muir sense of the word, give up the idea that wildlife can truly have self-sustaining populations without human intervention. For ferrets, the concept was already unrealistic given the problems with prairie dogs and plague, because vaccine or flea control would need to be continually applied. It was likely that in perpetuity, ferret populations would need continuous monitoring, vaccination, and restoration.

Edward Abbey wouldn't like Kansas, fragmented farmlands that are losing the last patches of native prairie to the bulldozer and plow, but Larry Haverfield's counterculture stance would make Ed Abbey and John Muir proud. His stubborn determination in the face of local government and popular opinion for the sake of the land, the ecosystem, is far more holistic and adaptable than any government bureaucracy. Private landowners own nearly twice as much rangeland (399 million acres) as the U.S. government, and typically it is the most fertile and suitable for sustainable prairie dog farming. Perhaps we no longer need to look to the remote corners of the globe for our modern-day visionaries and conservation heroes; they are fighting battles in nearby but unexpected places.

Epilogue

It is October and I return to UL Bend. It's the time of year when you wear insulated bib overalls and your hunter's cap into the field for the day and leave your beer in a cooler outside, without ice. You come back to camp after a day's work and the beer is the perfect temperature, going down easy as you grill dinner and drink can after can. I pull out my notebook and put pen to paper:

> Plump prairie dogs nibbling at nothing and digging at roots.
>
> A hard day's work on four-wheeler and foot, mapping the boundaries of prairie dog colonies: East Legg, North Dry Lake, East Dry Lake, North UL Bend Wilderness, Northeast UL Bend Wilderness, and two new colonies east of Legg Well.
>
> Does it get any better?
>
> Time has told me that what I once thought was an escape from real life is actually real life.
>
> Prairie dogs cry an alarm for three crows flying over, birds they rarely see compared to ferruginous hawks, prairie falcons and harriers; and that are only around now for the elk gut piles left down in the breaks by hunters.
>
> A day of overcast clouds in the west that made the whole afternoon look like sunset on the horizon. I find myself always looking west.

I watch the sunset shine soft golden light on the leaves of prairie grasses, and imagine them turning off the photosynthetic process for the day, halting energy production. I think deeper into the tissues and the green blood of chlorophyll that is structurally almost identical to human blood, only with the iron in hemoglobin replaced by magnesium in plants. I think of the chemical process going on deep in those cells, the process of eating light to produce energy, the fact that at dusk, photosynthesis reverses so plants emit small amounts of light.

I think of the first black-footed ferret I saw in the wild, just northwest of here, on a prairie dog colony not so very different from this one. The ferrets are long gone from there now, lost to the ravages of disease and politics. I remember the young man that I was back in 1999, rail-thin and fresh out of college. I do the math for the first time and realize that I have spent more than a decade of my life studying prairie dogs and their critically endangered predator. Feelings of mileage are mixed with enthusiasm, as I hope some ferrets still roam in these breaks above the Missouri, knowing that I will never tire of spending time in their presence—sleepless nights in the corners of prairie they inhabit.

Yet whenever I leave this place, even with the rise of conservation ethics and the invention of the practice of conservation biology in the past thirty years, I see increasing apathy about conserving such endangered species. Perhaps even you, the reader, are unsure of what to advocate for in this information age when the top news story changes every few minutes—global warming one day and elephant poaching the next. Do not worry, you are not alone. Even conservation biologists who guide the future of biodiversity at a global scale increasingly debate what to protect. Not only what to protect, but where, how, and when to let go. They debate about when to allow the extinction of a species and let economics tell us that it is simply too costly, too inconvenient to hold on—debates that threaten the pillars of the conservation ethic put forward by Emerson, Thoreau, Muir, Leopold, Soulé, and many others.

Allow me to explain. The simple answers to the question of what to protect, in a movement based on biodiversity, are species. These end units of evolution are the fundamental building blocks of conservation biology and its unit of currency. Yet, as science advances and species are described in more morphological and genetic detail, the separation of species begins to blur. As populations are fragmented by human development, individual populations frequently show some level of adaptation and genetic variation to their new island environment, leading to calls for protection of subgroups of species or populations. Other species like the red wolf and Florida panther hybridize, making it even more difficult to define what a unique species is and whether it deserves protection.

Even if we get over this taxonomic hurdle, there is debate about how to set priorities for protection. Do we protect all biodiversity? But that is impossible, right? There is only so much money to go around. Do we let some species go extinct and focus our limited resources on saving others? Do we let one species go for the sake of saving a dozen others from becoming endangered?

And when we select a species to restore, what are we trying to achieve? Do we have some ideal ecosystem in mind that we want to restore? To what point in time do we restore a Great Plains ecosystem—pre-1900, pre-colonial invasion, or the pre-human Pleistocene epoch?

Some ecologists say that restoration is futile, that from a long-term point of view an ecosystem is never in a stable state, always in flux, so it is ludicrous to think we should restore it to any one state. They argue that we need to let go of the past and accept a "new normal" because humans have created novel ecosystems. According to this view, we should focus our scientific efforts on understanding the processes at work in these new systems and refrain from attempts to reverse the direction in which humans are altering nature.

Yet, to really understand what to prioritize perhaps we need to go deeper. Although there are direct, utilitarian benefits to humankind from conserving biodiversity (think food, medicine, water quality, and so on), those who support conservation typically are motivated by something that is not as easily justified or given a dollar value: sentimentality. It is this human characteristic that is and always has been at the core of the conservation ethic and that links all conservation biologists as a community. A sentimental, altruistic bond that we find in common with other fields we hold dear, from global human rights, to national pride, and on down to small-town communities and even our own nuclear and extended families. We are simply enlarging the boundaries of the community we hold sentiment toward to include plants and animals, or as Aldo Leopold would say, "collectively: the land." It is perhaps the most selfless act to try to preserve something when it is not in the strictest sense physically, energetically, or evolutionarily beneficial to do so. We do it because something obvious or hidden deep in our psyches tells us that in the face of change it is right to keep something the way it is, return it to what it formerly was, or lift up something from the trouble it dwells in.

It is this type of sentimentality that is needed to preserve the black-footed ferret. It takes passion beyond the ordinary to restore this species back to the Great Plains, and the acts of a few can restore this species. Think of Bill Perry and his pioneering efforts to consolidate the Conata Basin into the single largest prairie dog complex on public land in the United States. Look to Larry Haverfield and Gordon Barnhardt, who have taken it upon themselves to create critical refuges in Kansas for native prairie species in a landscape of hate and paranoia.

At the time of writing this, despite releasing more than three thousand captive-reared kits across nineteen reintroduction sites over the past twenty-two years, we are technically only about one-quarter of the way to numerically recovering the black-footed ferret as a species. Today, if you visit Malta, Montana, you would never know ferrets had been raised there. Trailers have been sold to the highest bidder as surplus government property, the captive breeding pens have been bulldozed into the ground. Because of drought and overgrazing, the El Cuervo prairie dog colony has been reduced to a fraction of its former self, no longer the largest prairie dog population in the world. Elsewhere, prairie dog populations across the West continue to blink out of existence because of plague epizootics. Even where I stand right now, following the loss of more than half of the UL Bend prairie dog population to plague, the ferret population hangs on by a thin thread.

But there is hope. A plague vaccine is on the way. Financial incentive plans are being developed to entice land owners to tolerate prairie dogs. In Kansas, Larry Haverfield and his neighbors have won a legal decision that prohibits the county and state from forcing him to kill his prairie dogs. In the similarly restrictive State of Texas, a protected prairie dog population is being created in Caprock Canyon State Park. In Mexico, the Nature Conservancy bought the large El Uno Ranch that is adjacent to the El Cuervo prairie dog colony and began restoring prairie dogs, and even reintroduced bison in 2009. North of UL Bend, a group named the American Prairie Foundation has been buying ranches in an effort to create a three-million-acre prairie reserve in the center of Montana that can serve as America's version of the Serengeti, bringing attention to the Great Plains and its natural history, complete with restored populations of bison, swift foxes, prairie dogs, and ferrets.

But that is only what is happening now. It is still unclear what value we will put on wildlife over the next hundred or thousand years of human expansion. As we demand more from the planet, what will be the role of sentimentality in driving our future decisions to conserve and restore past diversity? What is clear is that today, in our lifetimes, the only way to recover the black-footed ferret is for individuals to care about them and the prairie dogs they rely on. Conservation of any species requires self-sacrifice by people who value another species above convenience and the economics of desire. For some, this can be thought of as stewardship, for others a duty to the animals. In the end, it is a very personal decision. I can only offer that for me, by preserving black-footed ferrets, we aren't just preserving a single animal or species; we are preserving an identity for the land. Even more, an identity for a country. A look at any map will show that the Great Plains dominates a great central swath of North America, yet the merits of its natural state have too often been overlooked for those of forests to the east and mountains to the west.

In the end, it is up to you to keep this book from becoming an elegy. Only with enough public voices will politicians, government agencies, and the rest of the world listen. Only through revolutionary thought will we be able to change laws and practices that continue to persecute declining prairie dog and black-footed ferret populations. Without action, this book will serve only as a memorial to efforts taken to recover ferrets and prairie dogs and the wild parts of the Great Plains as they once were.

Acknowledgments

Over a long term, hopefully humans will interact with black-footed ferrets for thousands of years into the future. During the brief period of time this book covers when the species was hopefully at its lowest, there have been hundreds of people who have helped conserve the species. I am but one biologist of many who have been part of the black-footed ferret recovery program, and of those hundreds of dedicated conservationists, there are many who have given much more of their lives to the black-footed ferret. It is each member of this conservation community whom I first acknowledge for their dedication and perseverance.

I particularly acknowledge two people who are featured in the book and have figured prominently in black-footed ferret conservation. Randy Matchett and Mike Lockhart have both been leaders in the conservation of the black-footed ferret for the past two decades. Although they will humbly shrug off such accolades, their vision to move conservation forward regardless of the personal sacrifice and their willingness to stand up and place their careers on the line to do what is right are what make them conservation visionaries.

I also acknowledge with gratitude the collaboration of Josh Millspaugh, Travis Livieri, Dean Biggins, and David Eads in my studies of ferret ecology. I particularly thank Josh for taking the risk of bringing me on as a graduate student with no funding and only the determination to work with black-footed ferrets and desire to learn more.

In composing the book, I thank Sarah Osmundson and the staff of the Phillips County Museum for help in researching the history of UL Bend, Montana. I am grateful for the help of my editor at University of California Press, Blake Edgar, and editorial advice from Merrik Bush-Pirkle and Pam Suwinsky, along with helpful comments on early drafts by Richard Reading and Mike Lockhart.

For me, writing has always been a solitary, private endeavor. The process itself is a necessary therapy and to think of publishing was rarely considered, not even a dream. For giving me the confidence to express myself and believe that what I put on the page is of value, I thank Mrs. Beth Ream, my high school writing instructor, who saw more potential in a shy, introspective boy than the boy saw in himself.

I thank my wife Cathy for her patience and support. Most families can persist with only one wildlife biologist, with the other spouse the more dependable and stay-at-home sort. She makes it work with two wildlife biologists in a household, and I am grateful for it.

I thank my father, Dr. Richard Jachowski, to whom this book is dedicated, for nurturing my desire to explore nature, from weekend expeditions in my youth into the salt marshes of the Chesapeake Bay to working with me in the Great Plains studying ferrets. I am also grateful for his willingness to act as my personal editor, having likely read every word I have ever written prior to it being published. The friendship we have built extends beyond genetic lines. Finally, I thank my mother for tolerating and nurturing my tendencies to travel, wander, and write about the outdoors.

Further Readings

Rather than provide an extensive bibliography, I have provided the following summary for those who are interested in reading in further detail about some of the topics covered in the chapters of this book. Most of the information on conservation of black-footed ferrets is contained in government reports or otherwise non-peer-reviewed literature that is poorly accessible to the public, yet there are still excellent resources readily available in the form of peer-reviewed scientific journal articles, books, and book chapters. Given their availability, I focus on the latter types of information in the following summaries, which are organized by readings relevant to the topic of each chapter. Each book title is followed by the publisher and year of publication. Each journal article title is followed by the journal name and volume number, the first and last page numbers of the article, and year of publication.

CHAPTER 1: PLEISTOCENE TO ANTHROPOCENE

A great starting point for learning to appreciate the Great Plains is the beautiful book recently compiled by Michael Forsberg entitled *Great Plains: America's Lingering Wild* (University of Chicago Press, 2009) that contains Michael's award-winning photography as well as the writing of Dan O'Brien and other prairie experts. For those interested in a more detailed history of the Great Plains, a good place to start is Richard Manning's *Grassland: The History, Biology, Politics and Promise of the American Prairie* (Penguin, 1997). David Wishart has put together the *Encyclopedia of the Great Plains,* a tremendous online resource that can be found at http://plainshumanities.unl.edu/encyclopedia/. For the central Montana area I focus on in this chapter, Richard Manning has written a book focusing on current restoration attempts to the area: *Rewilding the West: Restoration in a Prairie Landscape* (University of California Press, 2011).

Much of our historical knowledge of the ecology of the Great Plains began with the expedition of Meriwether Lewis and William Clark that was commissioned by then-president Thomas Jefferson in 1804. Lewis and Clark led the first modern exploration of the Great Plains by following the Missouri River from St. Louis to its source in the Rocky Mountains. Their expedition journals were compiled in *The Journals of Lewis and Clark* (Penguin, 2002). Their voyage is also popularly summarized in *Undaunted Courage* by Steven Ambrose (Simon and Schuster, 1997).

A review of the ecological role bison play is summarized by Alan Knapp and colleagues in "The Keystone Role of Bison in North American Tallgrass Prairie" (*BioScience* 49:39–50, 1999). A more general, yet nearly comprehensive review of the ecology of bison can be found in Dale Lott's book *American Bison: A Natural History* (University of California Press, 2002). Through the use of a variety of archeological techniques, anthropologists have attempted to re-create and describe historic bison migration routes. A key summary of these patterns has been described by Jeffrey Hanson in his article "Bison Ecology in the Northern Plains and a Reconstruction of Bison Patterns for the North Dakota Region" (*Plains Anthropologist* 29:93–113, 1984). An engaging account of current attempts to conserve bison on the Great Plains can be found in Dan O'Brien's *Buffalo for the Broken Heart* (Random House, 2002).

Further details on the natural history of black-footed ferrets and their first observations by humans can be found in "Paleobiology, Biogeography, and Systematics of the Black-Footed Ferret, *Mustela nigripes*" by Elaine Anderson and others (*Great Basin Naturalist Memoirs* 8:11–62, 1986); as well as "Fossils, Diet and Conservation of Black-Footed Ferrets" by Pamela Owen and others (*Journal of Mammalogy* 81:422–33, 2000).

Leading prairie dog expert John Hoogland summarized years of innovative research on the ecology of black-tailed prairie dogs in his book *The Black-Tailed Prairie Dog: Social Life of a Burrowing Mammal* (University of Chicago Press, 1995). Further details on the communication and social behavior of prairie dogs can be found in *Prairie Dogs: Communication and Community in an Animal Society* by Con Slobodichikoff and others (Harvard University Press, 2009). Additional information on prairie dogs and their associated prairie species is provided in *Prairie Dog Empire: A Saga of the Shortgrass Prairie* by Paul Johnsgard (University of Nebraska Press, 2005).

Early observations made by C. Hart Merriam on prairie dog distribution and behavior, as well as the roots of his thinking regarding the eventual justification of large-scale eradication campaigns, can be found in his early article "The Prairie Dog of the Great Plains" (*Yearbook of the Department of Agriculture* 257–70, 1902). Controversies surrounding the past extent of prairie dogs are presented by Lance Vermeire and colleagues in "The Prairie Dog Story: Do We Have It Right?" (*BioScience* 54:689–95, 2004) and a rebuttal by Steve Forrest in the article "Getting the Story Right: A Response to Vermeire and Colleagues" (*BioScience* 55:526–30, 2005). Finally, a collection of articles about past and current attempts to conserve black-tailed prairie dogs can be found in *Conservation of the Black-Tailed Prairie Dog* edited by John Hoogland (Island Press, 2006).

Prairie dog poisoning remains a common practice across the Great Plains. In the edited volume *Conservation of the Black-Tailed Prairie Dog* (Island Press, 2006), Steve Forrest and James Luchsinger provide an excellent summary of the historic and current state of prairie dog poisoning in their chapter "Past and Current Chemical Control of Prairie Dogs" (115–28). A short history of the extent of prairie dogs specific to the northern Great Plains can be found in David Roemer and Steve Forrest's article "Prairie Dog Poisoning in Northern Great Plains: An Analysis of Programs and Policies" (*Environmental Management* 20:349–59, 1996). A recent review of how widespread prairie dog poisoning remains and its part in limiting conservation success of prairie dogs and black-footed ferrets is summarized in "Challenges to Black-Footed Ferret Recovery: Protecting Prairie Dogs" (*Western North American Naturalist* 72:228–40, 2012) by Brian Miller and Richard Reading.

The increased risk of losing specialist or highly specialized species is a topic of increasing concern in conservation biology. To understand what is meant by *specialist* as opposed to *generalist species* requires knowledge of the concept of an ecological niche. An ecological niche is essentially a suite of attributes that encompasses all that a given species requires. Specialists have a fairly narrow range of attributes they can tolerate compared to generalists, which can persist with a wider range of attributes, or a wider niche. Peter Chesson provides a more thorough discussion on ecological niches in his article "Mechanism and Management of Species Diversity" (*Annual Review of Ecological Systematics* 31:343–66, 2000). Joanne Clavel and colleagues provide a good review of why specialists are at greatest risk, as well as the potential outcomes of losing specialists, in the article "Worldwide Decline of Specialist Species: Toward a Global Functional Homogenization?" (*Frontiers in Ecology and the Environment* 9:222–28, 2011).

The concept of bringing back species (or surrogates of species) not seen on the Great Plains or elsewhere in the western United States since the Pleistocene can be found in "Pleistocene Rewilding: An Optimistic Agenda for Twenty-First Century Conservation" by Josh Donlan and others (*American Naturalist* 168:660–81, 2006).

CHAPTER 2: DECLINE TOWARD EXTINCTION

Although much of the material discussed in this chapter regarding black-footed ferret population decline and subsequent conservation attempts in South Dakota during the 1950s through 1980s is from previously unpublished letters, notes, and government reports, there are some more widely available published sources. A short but engaging account of the decline of black-footed ferrets in South Dakota during the 1960s can be found in Faith McNulty's book *Must They Die?* (Doubleday and Company, 1970).

The first scientific reports on black-footed ferret ecology and behavior were published during this period, coinciding with the research by Con Hillman and others. Some of the first field observations are summarized by Con Hillman in "Field Observations of Black-Footed Ferrets in South Dakota" (*Transactions of the North American Wildlife and Natural Resources Conference* 33:433–43,

1968) and "Prairie Dog Distribution in Areas Inhabited by Black-Footed Ferrets" (*American Midland Naturalist* 102:185–87, 1979). Robert Sheets and colleagues published a series of ferret natural history findings that can be found in "Food Habits of the Black-Footed Ferret in South Dakota" (*Proceedings of the South Dakota Academy of Sciences* 48:58–61, 1969) and "Burrow Systems of Prairie Dogs in South Dakota" (*Journal of Mammalogy* 52:451–53, 1971).

Some of the findings from early attempts to transport ferrets to Patuxent Wildlife Research Center for captive breeding are summarized in articles by Con Hillman and Jim Carpenter entitled "Breeding Biology and Behaviour of Captive Black-Footed Ferrets" (*International Zoo Yearbook* 23:186–91, 1983) and "Husbandry, Reproduction, and Veterinary Care of Captive Ferrets" (*American Association of Zoo Veterinarians Annual Proceedings* 1977:36–47, 1978). Details on the loss of the first round of ferrets translocated from Mellette County, South Dakota, to Patuxent and their deaths following distemper inoculation are summarized in "Fatal Vaccine Induced Canine Distemper Virus Infection in Black-Footed Ferrets" (*Journal of the American Veterinary Medical Association* 169:961–64, 1976).

CHAPTER 3: REDISCOVERY

A detailed review of black-footed ferret rediscovery in Meeteetse, Wyoming, and subsequent management is provided in the book *Prairie Nights,* authored by Brian Miller, Richard Reading, and Steve Forrest (Smithsonian, 1996). Further details on the conflicts and discussions surrounding the management of the rediscovered ferret population can be found in *Conservation Biology and the Black-Footed Ferret* by Ulysses Seal and others (Yale University Press, 1989); *Averting Extinction: Reconstructing Endangered Species Recovery* by Tim Clark (Yale University Press, 2005); as well as the book chapter "The Black-Footed Ferret Recovery Program: Unmasking Professional and Organizational Weaknesses" by Richard Reading and Brian Miller (73–100 in the book *Endangered Species Recovery*, Island Press, 1994). A short summary of this period by Tim Clark can be found in "Black-Footed Ferret Recovery: A Progress Report" (*Conservation Biology* 1:8–13, 1987).

A great deal has been published in the scientific literature on black-footed ferret ecology as a result of studies that took place at Meeteetse, including multiple chapters in the edited volume entitled *The Black-Footed Ferret* (Great Basin Naturalist Memoirs, 1986). Details on demographic attributes of the Meeteetse ferret population can be found in Steve Forrest and colleagues' "Population Attributes for the Black-Footed Ferret at Meeteetse, Wyoming 1981–1985" (*Journal of Mammalogy* 69:261–73, 1988).

One of the primary tools wildlife biologists use to track wildlife is radiotelemetry, which involves the placement of a remote tracking device on an animal. Typically accomplished by implanting a transmitter inside the animal or attaching the transmitter onto the outside of an animal with a collar, this technology allows researchers to closely monitor movement and behavior of wildlife in a wild setting. A review of technology and quantitative approaches to analyzing the resulting data are provided in the book edited by Josh Millspaugh and John Marzluff entitled *Radio Tracking and Animal Populations* (Academic Press, 2001). A more condensed review of radio-telemetry and wildlife research and management is provided in the recent book chapter by Josh Millspaugh and colleagues entitled "Wildlife Radiotelemetry and Remote Monitoring" (258–83) in the *Wildlife Techniques Manual* edited by Nova Silvy (Johns Hopkins University Press, 2012).

Dean Biggins and his colleagues in particular did extensive research on black-footed ferrets involving radio-telemetry that has been published in numerous symposium proceedings and government reports, portions of which are also presented in peer-reviewed journals, including "Activity of Radio-Tagged Black-Footed Ferrets" (*Great Basin Naturalist Memoirs* 8:135–40, 1986) and "Black-Footed Ferret Activity during Late Summer and Fall at Meeteetse, Wyoming" (*Journal of Mammalogy* 92:705–9, 2011).

The use of passive integrated transponder (PIT) tags in marking ferrets is invaluable because of the tags' long life and small size. Widely used in domestic dogs and cats, these rice-grain-sized microchips are passive in that they do not emit a signal until a receiver comes into close contact. The receiver then sends a signal that is bounced back from the PIT tag to the receiver, indicating the nine-digit identification number of the implant. A review of their early use in ferrets is provided by Kathleen Fagerstone and Brad Johns's article, "Transponders as a Permanent Identification Markers for Domestic Ferrets, Black-Footed Ferrets, and Other Wildlife" (*Journal of Wildlife Management* 51:294–97, 1987).

Details on the ecology and theory behind mating tactics and polygamous behavior of otherwise typically solitary carnivores can be found in "The Mating Tactics and Spacing Patterns of Solitary Carnivores," a chapter by Mikael Sandell within the larger edited volume *Carnivore Behavior, Ecology and Evolution* (Comstock, 1989). This mating structure is common in ferrets and their taxonomic relatives, such as weasels. Because the latter have been studied in great detail, I highly recommend the book by Carolyn King and Roger Powell entitled *The Natural History of Weasels and Stoats: Ecology, Behavior, and Management* (Oxford University Press, 2007).

Many of the key concepts that led to the development of the field of conservation biology originated with concerns over the loss of species diversity. Key motivating factors were the observed loss of species and the advancement of the idea that smaller and more isolated islands of habitat had lower species diversity. This theory was developed and tested by E. O. Wilson and Robert MacArthur in their book *The Theory of Island Biogeography* (Princeton University Press, 1967), and was expanded beyond islands to apply to habitat fragmentation, where, as habitat degradation increased, species loss would similarly increase. A highly approachable overview of this theory and a gentle introduction to the field of conservation biology are provided in David Quammen's excellent book *Song of the Dodo* (Scribner, 1997).

The tenets of the field of conservation biology are covered in Michael Soule's influential article "What Is Conservation Biology?" (*BioScience* 35:727–34, 1985) and have subsequently been reviewed in multiple college textbooks; one of the best is *Principles of Conservation Biology* (Sinaeur, 2005). Beyond scientific interest, the field of conservation biology is a blending of natural and social sciences and has been defined as a mission-oriented discipline to conserve biodiversity. Although scientists were slow to take on the role of advocates, one of the most powerful arguments was put forward by Daniel Janzen in his revolutionary article "The Future of Tropical Ecology" (*Annual Review of Ecology and Systematics* 17:305–24, 1986), in which he contends that ecologists need to advocate for the conservation of the systems they study rather than being passive observers.

To protect biodiversity, conservation biologists often set thresholds for when species are at greatest risk of extirpation or extinction. These minimum viable population sizes are based on the early mathematical models developed by Mark Shaffer on grizzly bears in the Yellowstone ecosystem and have been refined over time to allow managers to predict the outcome of management actions on the viability of populations. Termed "population viability analysis," a thorough review of the topic is covered in a book by the same name, *Population Viability Analysis* (University of Chicago Press, 2002) edited by Steven Beissinger and Dale McCullough.

The Endangered Species Preservation Act was passed by Congress in 1966 to allow for listing of U.S. species, eventually including fourteen mammals, including the black-footed ferret; thirty-six birds; six reptiles and amphibians; and twenty-two fishes. The law was amended in 1969 to extend to species worldwide and was rewritten into its present form as the Endangered Species Act in 1973 under then-president Richard Nixon. Details on the Endangered Species Act (ESA) and species listed or petitioned for protection can be found on the U.S. Fish and Wildlife Service's website: www.fws.gov/endangered/. The process of listing species typically requires public petitioning for a species to be added to the list, and a thorough review of its status, but the process can be biased toward specific types of species. A number of critical reviews of the ESA have been produced that relate to biases and trends in listing decisions, such as Marco Restani and John Marzluff's article "Funding Extinction? Biological Needs and Political Realities in the Allocation of Resources to Endangered Species Recovery" (*BioScience* 52:169–77, 2002). Once a species is listed, by law critical habitat should be designated and a recovery plan developed. However, even once developed, these two components can be highly contentious. Alan Clark and colleagues put together an excellent short review of ESA recovery planning in "Improving U.S. Endangered Species Act Recovery Plans: Key Findings and Recommendations of the SCB Recovery Plan Project" (*Conservation Biology* 16:1510–19, 2002). Finally, while some citizens and lawmakers use evidence of a slow rate of successfully recovering and delisting species as justification for the abandonment of the ESA, conservation advocates blame delays in the initial listing of species for the current slow rate of success. A review of this contentious issue was summarized in an article by Martin Taylor and colleagues, "The Effectiveness of the Endangered Species Act: A Quantitative Analysis" (*BioScience* 55:360–68, 2005).

The National Forest Management Act, passed in 1976, revolutionized the management of U.S. Forest Service lands by requiring that U.S. Forest Service land managers evaluate multiple alternatives to management beyond just resource extraction. It included the prioritization of management decisions for wildlife and fish species and provided an avenue for the protection of species and ecosystems on federal lands over economic gain alone. This type of multiple-use land management (that originated in the 1960s with the Multiple-Use and Sustained-Yield Act) has served as a guiding paradigm across federally owned and managed U.S. Forest Service lands over the past several decades. Some of the benefits of this type of management approach are profiled in the article by Dan Wilcove entitled "Protecting Biodiversity in Multiple-Use Lands: Lessons from the US Forest Service" (*Trends in Ecology and Evolution* 4:385–88, 1989).

CHAPTER 4: CAPTIVE BREEDING

Following rediscovery, careful advances were made in captive breeding and husbandry techniques. These steps are reported by Brian Miller and colleagues in the book *Prairie Night* (Smithsonian, 1996) as well as in a series of scientific journal articles, including "A Comparison between Black-Footed Ferrets, Domestic Ferrets, and Siberian Polecats" (*Zoo Biology* 9:201–10, 1990); "Biology of the Endangered Black-Footed Ferret and the Role of Captive Propagation in Conservation" (*Canadian Journal of Zoology* 66:765–73, 1988); and "Rehabilitation of a Species: The Black-Footed Ferret" (*Wildlife Rehabilitation* 9:183–92, 1992). Beth Williams and colleagues also documented reproductive ecology of black-footed ferrets in the article "Reproductive Biology and Management of Captive Black-Footed Ferrets" (*Zoo Biology* 10:383–98, 1991). Early successes in black-footed ferret captive breeding research are summarized by Astrid Vargas and colleagues in "Growth and Development of Captive-Raised Black-Footed Ferrets" (*American Midland Naturalist* 135:43–52, 1996).

In addition to government facilities, zoological parks and nongovernmental research institutions have played a key role in black-footed ferret captive breeding advancements. Jo Gayle Howard of the Smithsonian Institution deserves particular attention for her work with captive ferret breeding and artificial insemination, a portion of which is reported in "Age-Dependent Changes in Sperm Production, Semen Quality, and Testicular Volume in the Black-Footed Ferret" (*Biology of Reproduction* 63:179–87, 2000); "Sperm Viability in the Black-Footed Ferret Is Influenced by Seminal and Medium Osmolality" (*Cryobiology* 53:37–50, 2006); "Environment Influences Morphology and Development for in situ and ex situ Populations of Black-Footed Ferrets" (*Animal Conservation* 8:321–28, 2005); "Approaches and Efficacy of Artificial Insemination in Felids and Mustelids" (*Theriogenology* 71:130–48, 2009).

Outside of the laboratory, Dean Biggins and colleagues did extensive research on the value of preconditioning ferrets prior to release, some of which is reported in "The Effect of Rearing Method on Survival of Reintroduced Black-Footed Ferrets" (*Journal of Wildlife Management* 62:643–53, 1998); "Influence of Prerelease Experience on Reintroduced Black-Footed Ferrets" (*Biological Conservation* 89:121–29, 1999); and "Black-Footed Ferrets and Siberian Polecats as Ecological Surrogates and Biological Equivalents" (*Journal of Mammalogy* 92:710–20, 2011).

CHAPTER 5: FALL

Judy Blunt wrote an entertaining memoir about life on a ranch just north of UL Bend National Wildlife Refuge in southern Phillips County, Montana, entitled *Breaking Clean* (Knopf, 2002) that provides an introduction to the history, isolation, and lifestyle of cattle ranchers in the area. From a wildlife conservation perspective, Richard Manning has put together a summary of current conservation actions in central Montana entitled *Rewilding the West: Restoration in a Prairie Landscape* (University of California Press, 2011).

Animal migrations, particularly those of long distance by large herbivores, are of particular conservation interest at the moment, and for good reason! These behaviors capture the imagination of the public but are increasingly imperiled by human disturbance. Joel Berger does an excellent job of reviewing the most heavily affected herbivores in his article "The Last Mile: How to Sustain Long-Distance Migration in Mammals" (*Conservation Biology* 18:320–31, 2004). Information on the migratory patterns of pronghorn of the Great Plains and how these behaviors are affected by human disturbance are reviewed in the book chapter by Cormack Gates and colleagues entitled "The Influence of Land Use and Fences on Habitat Effectiveness, Movements, and Dispersal of Pronghorn in the Grasslands of North America" (277–94) in *Fencing for Conservation* (Springer, 2012) edited by Mark Somers and Matt Hayward. A general review of pronghorn ecology can be found in the book by Richard McCabe and others entitled *Prairie Ghost: Pronghorn and Human Interaction in Early America* (University Press of Colorado, 2004). Further, John Byers has written an engaging account of the natural history of pronghorn in *Built for Speed: A Year in the Life of Pronghorn* (Harvard University Press, 2003).

Reviews of local landowner attitudes toward ferret reintroduction in southern Phillips County, Montana, can be found in Richard Reading and Steven Kellert's article "Attitudes toward a Proposed Reintroduction of Black-Footed Ferrets" (*Conservation Biology* 7:569–80, 1993). A broad review of human attitudes toward prairie dogs is provided in the chapter by Berton Lee and others entitled "Attitudes and Perceptions about Prairie Dogs" (108–14) in the book *Conservation of the Black-Tailed Prairie Dog* (Island Press, 2006).

A great deal of research and information exists on the potential conflicts between prairie dogs and livestock producers because of competition for forage. Many livestock producers argue that as a result of prairie dog clipping of vegetation, there is a clear reduction in forage available to their cattle. By contrast, conservationists often point to the beneficial aspects of prairie dogs, such as plant clipping enriching nutrient content, their digging and burrow systems aerating the soil, and other lines of evidence (see following). In truth, the extent of this issue is often site and scale dependent. An excellent place to begin learning more about what researchers have found is in James Detling's chapter "Do Prairie Dogs Compete with Livestock?" (65–88) in the book *Conservation of the Black-Tailed Prairie Dog* (Island Press, 2007) edited by John Hoogland.

The ecological role of prairie dogs as ecosystem engineers is widely appreciated by scientists because of the impact of their burrow systems and herbivory on structure ecosystems. Multiple studies from across the range of the prairie dog support the concept that prairie dog colonies increase habitat heterogeneity and species diversity. Some of these findings can be found in research on black-tailed prairie dog colonies in Kansas by Ron VanNimwegen and colleagues in "Ecosystem Engineering by a Colonial Rodent: How Prairie Dogs Structure Rodent Communities" (*Ecology* 89:3298–3305, 2008) and in Mexico by Gerardo Ceballos and others in "Influence of Prairie Dog (*Cynomys ludovicianus*) on Habitat Heterogeneity and Mammalian Diversity in Mexico" (*Journal of Arid Environments* 41:161–72, 1999). Evidence for the role of Gunnison's prairie dog as ecosystem engineer can be found in the research by Randy Bangert and Con Slobodchikoff in "Conservation of Prairie Dog Ecosystem Engineering May Support Arthropod Beta and Gamma Diversity" (*Journal of Arid Environments* 67:100–115, 2006) and "The Gunnison's Prairie Dog Structures a High Desert Grassland Landscape as a Keystone Engineer" (*Journal of Arid Environments* 46:357–69, 2000). The broader global importance of social, burrowing rodents on the world's grasslands is profiled in the recent article by Ana Davidson and others entitled "Ecological Roles and Conservation Challenges of Social, Burrowing, Herbivorous Mammals in the World's Grasslands" (*Frontiers in Ecology and the Environment* 10:477–86, 2012).

Grasslands bird diversity is often higher on prairie dog colonies than surrounding areas, as discussed in the article by William Agnew and others entitled "Flora and Fauna Associated with Prairie Dog Colonies and Adjacent Ungrazed Mixed Grasslands in Western South Dakota" (*Journal of Range Management* 39:135–39, 1986). A variety of prairie specialists including horned larks, burrowing owls, ferruginous hawks, and mountain plovers are most often sighted on or near prairie dog colonies. The mountain plover deserves particular attention because of its declining status and the fact that it is the most closely linked to prairie dogs of all avian species. In central Montana, Steve Dinsmore has spent more than a decade studying mountain plovers and documenting how their nesting habitats and behavior are closely linked to the persistence of prairie dog populations. Some of this work can be found in "Mountain Plover Population Responses to Black-Tailed Prairie Dogs in Montana" (*Journal of Wildlife Management* 69:1546–53, 2005) and "Responses of Mountain Plovers to Plague-Driven Dynamics of Black-Tailed Prairie Dog Colonies" (*Landscape Ecology* 23:689–97, 2008).

The leading source of information on details of prairie dog ecology ranging from the complexity of their burrow systems to social structure is John Hoogland's book *The Black-Tailed Prairie Dog: Social Life of a Burrowing Mammal* (University of Chicago Press, 1995). Further information on prairie dog communication and social structure, and Gunnison's prairie dog species in particular, is provided in Con Slobodchikoff and colleagues' excellent book *Prairie Dogs: Communication and Community in an Animal Society* (Harvard University Press, 2009).

CHAPTER 6: WINTER

Although few people have attempted to track black-footed ferrets over winter, and even fewer people have attempted to write about their efforts, some information can be found in "Winter Ecology of the Black-Footed Ferret at Meeteetse, Wyoming" by Louis Richardson and colleagues (*American Midland Naturalist* 117:225–39, 1987).

The work of Olaus Murie published in 1951 entitled *The Elk of North America* (Wildlife Management Institute, 1951) still serves as a benchmark publication on the species and in the broader wildlife biology literature. A more recent update authored by elk experts Jack Ward Thomas and Dale E. Toweill has been published under the same name (*Elk of North America,* Wildlife Management Institute, 1982). Specific to central Montana, Richard Mackie authored an early monograph on elk and other larger herbivores in "Range Ecology and Relations of the Mule Deer, Elk and Cattle in the Missouri River Breaks, Montana" (*Wildlife Monographs* 20:3–79, 1970). Bison in particular have been missing from central Montana prairies, yet there are ongoing efforts to restore the species by the American Prairie Foundation that are profiled in the book *Rewilding the West: Restoration in a Prairie Landscape* (University of California Press, 2011) by Richard Manning.

Predators are known to have a large impact on black-footed ferret survival, causing up to 95 percent of known black-footed ferret mortalities. As discussed in the suggested readings for chapter 4, preconditioning of captive ferrets to predators prior to release was at one time a common practice to increase ferret survival post-release. To further limit mortality of released ferrets to terrestrial predators, electric predator fencing has been used to try to limit the loss of ferrets to coyotes and badgers. Stuart Breck and colleagues critically evaluated the effectiveness of predator fencing and found no beneficial effects of fencing on black-footed ferret survival. These findings are reported in "Does Predator Management Enhance Survival of Reintroduced Black-Footed Ferrets?" (203–9), published as part of *Recovery of the Black-Footed Ferret—Progress and Continuing Challenges* (U.S. Geological Survey Scientific Investigations Report, 5293, 2005). More recently, Sharon Poessel and others have hypothesized that owl predation could significantly reduce ferret survival in the article "Landscape Features Influence Postrelease Predation on Black-Footed Ferrets" (*Journal of Mammalogy* 92:732–41, 2011).

Detailed information on the reconstructed path plague took in spreading from the West to the East, starting in California, is provided by Jennifer Adjemian and colleagues in "Initiation and Spread of Traveling Waves of Plague, *Yersinia pestis,* in the Western United States" (*American Journal of Tropical Medicine and Hygiene* 76:365–75, 2007). The early report mentioned in the chapter regarding plague and black-footed ferrets published by Mike Antolin and others can be found in "The Influence of Sylvatic Plague on North American Wildlife at the Landscape Level, with Special Emphasis on Black-Footed Ferret and Prairie Dog Conservation" (*Transactions of the North American Wildlife and Natural Resources Conference* 67:104–27, 2002). An excellent review of the history and ecology of the disease sylvatic plague is provided in the article "Natural History of Plague: Perspectives from More Than a Century of Research" (*Annual Review of Entomology* 50:505–28, 2005) by Ken Gage and Mike Kosoy.

Research by Brian Holmes on trapping and sampling of small mammals for fleas on Charles M. Russell National Wildlife Refuge and subsequently testing those fleas for plague bacteria is summarized in his article "No Evidence of Persistent Yersinia pestis Infection at Prairie Dog Colonies in North-Central Montana" (*Journal of Wildlife Diseases* 42:164–66, 2006). The work by Dave Hanson regarding plague-positive fleas identified on Fort Belknap Indian Reservation is detailed in his article "High Prevalence of Yersinia pestis in Black-Tailed Prairie Dog Colonies during an Apparent Enzootic Phase of Sylvatic Plague"(*Conservation Genetics* 8:789–95, 2007). Outside of Montana, a great deal of work has been done on the potential for small rodents (other than prairie dogs) to transmit or act as reservoirs for the disease. Some of this work is presented in the papers by Paul Stapp and Daniel Salkeld entitled "Evidence for the Involvement of an Alternative Rodent Host in the Dynamics of Introduced Plague in Prairie Dogs" (*Journal of Animal Ecology* 78:807–17, 2009) and "Plague Outbreaks in Prairie Dog Populations Explained by Percolation Thresholds of Alternative Host Abundance" (*Proceedings from the National Academy of Sciences* 107:14247–250, 2010).

Findings from Brendan Moynahan's long-term monitoring of the central Montana sage grouse population can be found summarized in "Factors Contributing to Process Variance in Annual Survival of Female Greater Sage-Grouse in Montana" (*Ecological Applications* 16:1529–38, 2006). Sage grouse conservation is a topic of increasing interest, and overwintering areas are of key importance in mitigating the impact of energy development. Some of the research and guidelines for managing winter sage grouse habitat can be found in Kevin Doherty's article "Greater Sage-Grouse Winter Habitat Selection and Energy Development" (*Journal of Wildlife Management* 72:187–95, 2008).

CHAPTER 7: SPRING

Although scent marking in black-footed ferrets is not well understood, there has been some discussion of winter scent marking by ferrets documented by Louis Richardson in "Winter Ecology of Black-Footed Ferrets at Meeteetse, Wyoming" (*American Midland Naturalist* 117:225–39, 1987). By contrast, there is a great deal known about scent marking and territorial behavior of weasels and stoats (the close taxonomic relatives of ferrets) that could be of interest. In particular, Carolyn King and Roger Powell provide an excellent review of the behavioral ecology of weasels and stoats in *The Natural History of Weasels and Stoats: Ecology, Behavior, and Management* (Oxford University Press, 2007), including a discussion of scent marking.

A review of lek breeding behavior by avian species can be found in most textbooks on ornithology. Essentially involving the aggregation of males and females on a regularly used display area where mate selection and breeding occurs, these lekking grounds provide the ideal opportunity for biologists and managers to annually count grouse population sizes. The number they count during each spring survey of a lek is used by state and federal agency biologists as an index of the size of the population, and comparisons of that index over time allow for rough assessments of how the populations are doing. However, some limitations exist because of surveyor error and biases when counting, and there are a number of critical reviews, including "Evaluation of Lek-Count Index for Greater Sage Grouse" (*Wildlife Society Bulletin* 32:56–68, 2004) by Daniel Walsh and colleagues.

A great deal has been published on sage grouse over the past decade because of conservation concerns. A good synthesis of our current knowledge about the species was compiled by Steven Knick and John Connelly in *Greater Sage Grouse: Ecology and Conservation of a Landscape Species and Its Habitats* (University of California Press, 2011). Specific to central Montana, findings of Brendan Moynahan's research on sage grouse that were mentioned in this chapter can be found in "Factors Affecting Nest Survival of Greater Sage-Grouse in Northcentral Montana" (*Journal of Wildlife Management* 71:1773–83, 2007).

Details on the impact of West Nile virus on avian taxa can be found in work synthesized by Sarah Wheeler entitled "Differential Impact of West Nile Virus on California Birds" (*Condor* 111, 1–20). An early summary of the extent of knowledge on the impact of West Nile virus on sage grouse can be found in the work of Dave Naugle and colleagues in their articles "West Nile Virus and Sage-Grouse: What More Have We Learned?" (*Wildlife Society Bulletin* 33, 616–23, 2005) and "West Nile Virus: Pending Crisis for Greater Sage-Grouse" (*Ecology Letters* 7:704–13, 2004). A recent synthesis of the impact of energy development on sage grouse and other wildlife in the Great Plains is contained in the edited volume by Dave Naugle entitled *Energy Development and Wildlife Conservation in Western North America* (Island Press, 2011).

Prairie dog communication is quite complex, with a wide range of calls that are able to match with specific visual cues. Details on the fascinating research Con Slobodchikoff and others have done on prairie dog communication can be found in the book *Prairie Dogs: Communication and Community in an Animal Society* (Harvard University Press, 2009). The results of more than thirty-one years of research by John Hoogland on prairie dog behavior and social systems are summarized in *The Black-Tailed Prairie Dog: Social Life of a Burrowing Mammal* (University of Chicago Press, 1995) and multiple papers, including a recent article "Prairie Dogs Disperse When All Close Kin Have Disappeared" (*Science* 339, 1205–7, 2013).

Black-footed ferrets exhibit a "fast" life history strategy in that they live for a relatively short period of time and become reproductively mature at a relatively young age. Currently there are very few studies that have attempted to quantify these life history traits, in large part because of the difficultly in monitoring ferret populations. One source of valuable information was presented by Martin Grenier and colleagues following monitoring of the reintroduced Shirley Basin population, where they observed "Rapid Population Growth of a Critically Endangered Carnivore" (*Science* 317:219, 2007). Their findings further support the concept that black-footed ferrets can quickly increase in number when adequate habitat is provided.

CHAPTER 8: SUMMER

A number of predictive mathematical models have been developed to assess the ability of a specific reintroduction to sustain a ferret population and to predict how many ferrets a site could potentially maintain. The most widely used model, first developed by Dean Biggins and colleagues in 1993, was based on the energetic demands of ferrets. In it, the number of ferrets a site could maintain was roughly a function of the abundance of prairie dogs at that site and the number of prairie dogs need to sustain a ferret family over the course of a year. Dean and coauthors published this model in the article "A Technique for Evaluating Ferret Habitat" (*U.S. Fish and Wildlife Service Biological Report* 13:73–88, 1993). Dean and others then further revised the model and published a modification accounting for territoriality in the 2006 article "Evaluating Habitat for Black-Footed Ferrets: Revision of an Existing Model" (*U.S. Geological Survey Scientific Investigations Report* 5293:143–50, 2005). Official down-listing criteria of black-footed ferrets under the Endangered Species Act requires ten populations of at least thirty breeding adults. However, to many conservation biologists, the number of thirty is too low. The 50/500 rule was proposed by Ian Franklin in the pioneering article "Evolutionary Change in Small Populations" in *Conservation Biology*: an evolutionary-ecological perspective (Sinauer, 1980) provides a still generally accepted rule of thumb that populations need to contain at least fifty individuals to avoid negative consequences of inbreeding, and at least five hundred individuals to sustain enough genetic diversity to maintain adaptive potential to change.

Prairie dog translocation is an increasingly common practice involving the planned trapping, transport, and release of prairie dogs onto another site for conservation purposes. Although this process sounds simple, trapping and moving prairie dogs is time consuming, and survival rates for translocated prairie dogs can be quite low. A good review of the practice is provided by Dustin Long and others in "Establishment of New Prairie Dog Colonies by Translocation" (188–209) in *Conservation of the Black-Tailed Prairie Dog* (Island Press, 2007) edited by John Hoogland.

Further details on some of the research I conducted with colleagues at UL Bend on black-footed ferret spatial ecology and behavior is published in "Implications of Black-Tailed Prairie Dog Spatial Dynamics to Black-Footed Ferrets" (*Natural Areas Journal* 28:14–25, 2008) and "Resource Selection by Black-Footed Ferrets in South Dakota and Montana" (*Natural Areas Journal* 31:218–25, 2011).

A general review by Tim Vosburg and colleagues of the practice of prairie dog recreational shooting can be found in the book *Conservation of the Black-Tailed Prairie Dog* (Island Press, 2006). In addition to potential direct effects on prairie dog density and abundance, some of the detrimental effects of recreational shooting on prairie dogs and the wider ecological community are reviewed in the work of Jonathan Pauli and Steven Buskirk in "Risk-Disturbance Overrides Density Dependence in a Hunted Colonial Rodent, the Black-Tailed Prairie Dog" (*Journal of Applied Ecology* 44:1219–30, 2007). Jonathan and Steven also bring up potentially important eco-toxicological implications of prairie dog shooting in "Recreational Shooting of Prairie Dogs: A Portal for Lead Entering Wildlife Food Chains" (*Journal of Wildlife Management* 71:103–8, 2007). An interesting account of the combined effect of recreational shooting and the presence of black-footed ferrets on prairie dogs is reported by Dean Biggins and colleagues in "Black-Footed Ferrets and Recreational Shooting Influences the Attributes of Black-Tailed Prairie Dog Burrows" (*Western North American Naturalist* 72:158–71, 2012).

CHAPTER 9: CHIHUAHUA

A great deal of research on the prairie ecosystem and the prairie dog populations surrounding Janos, Mexico, has been carried out by Geraldo Ceballos, Rurik List, and others at the Instituto de Ecologia at the Universidad Nacional Autónoma de Mexico, including "Factors Associated with Long-Term Changes in Distribution of Black-Tailed Prairie Dogs in Northwestern Mexico" (*Biological Conservation* 145:54–61, 2011); "Avian Diversity in a Priority Area for Conservation in North America: The Janos-Casas Grandes Prairie Dog Complex and Adjacent Habitats in Northwestern Mexico" (*Biodiversity and Conservation* 15:3801–25, 2006); *Biodiversity, Ecosystems and Conservation in Northern Mexico* (Oxford University Press, 2005); "Rapid Response of a Grassland Ecosystem to an Experimental Manipulation of a Keystone Rodent and Domestic Livestock" (*Ecology* 91:3189–200, 2010). The conservation status of the area has similarly been reviewed by this group of scientists in a series of articles, including "Rapid Decline of a Grassland System and Its Ecological and Conservation Implications" (*PLoS ONE* 5:e8562, 2010); "Historic Distribution and Challenges to Bison Recovery in the Northern Chihuahua Desert" (*Conservation Biology* 21:1487–94, 2007); and "The Janos Biosphere Reserve" (*International Journal of Wilderness* 16:35–41, 2010).

A broad review of the process and status of black-footed ferret reintroduction can be found in an article I authored with Mike Lockhart entitled "Reintroducing the Black-Footed Ferret to the Great Plains of North America" (*Small Carnivore Conservation* 41: 58–64, 2009). In 2007, Martin Grenier and colleagues published dramatic documentation of delayed black-footed ferret reintroduction success at Shirley Basin, Wyoming, on surprisingly large white-tailed prairie dog colonies following initial releases in the early 1990s, in the article "Rapid Population Growth of a Critically Endangered Carnivore" (*Science* 317:779, 2007). These findings supported the idea that large populations of prairie dogs are needed to conserve and restore black-footed ferrets and were similarly supported by my then comprehensive review of the reintroduction program published in "The Importance of Thinking Big: Large-Scale Prey Conservation Drives Black-Footed Ferret Reintroduction Success" (*Biological Conservation* 144:1560–66, 2011).

Details on the use of spotlighting to find and monitor ferrets, as well as drawings and descriptions of the traps used to capture ferrets, are provided by Dean Biggins and others in "Monitoring Black-Footed Ferrets during Reestablishment of Free-Ranging Populations: Discussion of Alternative Methods and Recommended Minimum Standards" (*U.S. Geological Survey Scientific Investigations* 5293:155–74, 2005).

Although much attention has been paid to the impact of sylvatic plague on black-footed ferrets, canine distemper virus has also had a limiting impact on conservation of the species. Impacts began with the last known wild ferret populations at Mellette County, South Dakota, and Meeteetse, Wyoming, which are summarized by Tom Thorne and Elizabeth Williams in "Disease and Endangered Species: The Black-Footed Ferret as a Recent Example" (*Conservation Biology* 2:66–74, 1988) and "Canine Distemper in Black-Footed Ferrets from Wyoming" (*Journal of Wildlife Diseases* 3:385–98, 1988).

Following the collapse of a population, the remaining small group of founding individuals is prone to a number of genetic problems, including the loss of genetic diversity that could harm subsequent fitness if and when the population recovers. A good resource on these conservation-related genetic issues is provided by Richard Frankham and others in the book *Introduction to Conservation Genetics* (Cambridge University Press, 2010). Early work by Sam Wisely and colleagues on the genetic history of black-footed ferrets can be found in "Genetic Diversity and Fitness in Black-Footed Ferrets before and during a Bottleneck" (*Journal of Heredity* 93:231–37, 2002) and "Evaluation of the Genetic Management of the Endangered Black-Footed Ferret" (*Zoo Biology* 22:287–98, 2003). Results from the biomedical survey discussed in this chapter that was conducted by Sam Wisely and colleagues can be found in "Genotypic and Phenotypic Consequences of Reintroduction History in the Black-Footed Ferret" (*Conservation Genetics* 9:389–99, 2008) and "An Unidentified Filarial Species and Its Impact on Fitness in Wild Populations of the Black-Footed Ferret" (*Journal of Wildlife Disease* 44:53–64, 2008).

CHAPTER 10: CONATA BASIN

Information on the Battle of Wounded Knee was powerfully described by Peter Matthiessen in his book *In the Spirit of Crazy Horse* (Penguin, 1992). A more recent account of life on the Pine Ridge Reservation can be found in Ian Frazier's book *Life on the Rez* (Picador, 2001).

The process of mapping the location of prairie dog burrows in Conata Basin, and resulting maps, are published in my article "Implications of Black-Tailed Prairie Dog Spatial Dynamics to Black-Footed Ferrets" (*Natural Areas Journal* 28:14–25, 2008). Details on the underground structure of prairie dog burrows and the location of black-footed ferret den sites were reported by Robert Sheets and others in "Burrow Systems of Prairie Dogs in South Dakota" (*Journal of Mammalogy* 52:451–53, 1971) and "Food Habits of the Black-Footed Ferret in South Dakota" (*Proceedings of the South Dakota Academy of Sciences* 48:58–61, 1969). Details on my use of a burrow den camera to film and monitor the behavior of wild female black-footed ferrets and their litters can be found in "Notes on Black-Footed Ferret Detectability and Behavior" (*Prairie Naturalist* 39:99–104, 2007).

A majority of published research on black-footed ferret behavior and ecology during the past two decades has taken place in the Conata Basin of South Dakota, where Travis Livieri has led black-footed ferret monitoring efforts in collaboration with research by Dean Biggins, David Eads, myself, and others. This work has resulted in findings on a wide variety of topics related to black-footed ferret behavior and conservation, a sample of which can be found in the articles "Black-Footed Ferret Digging Activity in Summer" (*Western North American Naturalist* 72:140–47, 2012) and "Importance of Lunar and Temporal Conditions for Spotlight Surveys of Adult Black-Footed Ferrets" (*Western North American Naturalist* 72:179–90, 2012). Findings on black-footed ferret space use can be found in the articles "Home-Range Size and the Spatial Organization of Black-Footed Ferrets in South Dakota" (*Wildlife Biology* 16:66–76,

2010) and "Black-Footed Ferret Home Ranges in Conata Basin, South Dakota" (*Western North American Naturalist* 72:196–205, 2012). Findings related to black-footed ferret resource selection were reported in the articles "Resource Selection by Black-Footed Ferrets in South Dakota and Montana" (*Natural Areas Journal* 31:218–25, 2011) and "Post-Breeding Resource Selection by Adult Black-Footed Ferrets in Conata Basin, South Dakota" (*Journal of Mammalogy* 92:760–70, 2011). Findings related to the efficacy of using translocation of wild adults as a management tool to initiate populations are described in the article "Movements and Survival of Black-Footed Ferrets Associated with an Experimental Translocation in South Dakota" (*Journal of Mammalogy* 92:742–50, 2011). Cynthia Cain and colleagues have conducted work documenting the genetic diversity of the Conata Basin population over time in "Genetic Evaluation of a Reintroduced Population of Black-Footed Ferrets" (*Journal of Mammalogy* 92:751–59, 2011). Sharon Poessel and colleagues published work on predation risk and ferrets of Conata Basin in "Landscape Features Influence Post-Release Predation on Endangered Black-Footed Ferrets" (*Journal of Mammalogy* 92:732–41, 2011).

The close temporary association between coyotes and badgers has long been observed by naturalists, and ecologist Steven Minta and colleagues have documented how coyotes in particular benefit from this pairing in their article "Hunting Associations between Badgers and Coyotes" (*Journal of Mammalogy* 73:814–20, 1992).

The relationship between food availability and prairie dog population dynamics has been the subject of investigation across multiple populations and Western states. The reference to prairie dog colony expansion, density, and forage availability in this chapter has most recently been shown experimentally in Canada by Natasha Lloyd and colleagues in "Food Limitation at Species Range Limits: Impacts of Food Availability on the Density and Colony Expansion of Prairie Dog Populations at their Northern Periphery" (*Biological Conservation* 161, 110–37, 2013).

CHAPTER 11: PLAGUE

An excellent review of plague can be found in the article "Natural History of Plague: Perspectives from More Than a Century of Research" (*Annual Review of Entomology* 50:505–28, 2005) by Ken Gage and Mike Kosoy. An introduction to the impact of plague on North American wildlife can be found in Dean Biggins and Mike Kosoy's article, "Influence of Introduced Plague on North American Mammals: Implications from Ecology of Plague in Asia" (*Journal of Mammalogy* 82:906–16, 2001). A summary of frontiers in plague research in North America can be found in the forum article by Rebecca Eisen and colleagues from the Centers for Disease Control entitled "Studies of Vector Competency and Efficiency of North America Fleas for Yersinia pestis: State of the Field and Future Research Needs" (*Journal of Medical Entomology* 46:737–44, 2009). In addition, a recent special edition of the journal *Vector-borne and Zoonotic Disease* (10, 2010) synthesizes many important recent findings regarding plague and particularly its impact on prairie dog and black-footed ferret populations.

The impact of plague on prairie dogs can be both devastating and dynamic, as reported in multiple articles, such as "Interspecific Comparison of Sylvatic Plague in Prairie Dogs" (*Journal of Mammalogy* 82:894–905, 2001); "A Plague Epizootic in the Black-Tailed Prairie Dog" (*Journal of Wildlife Disease* 42:74–80, 2006); and "Spatiotemporal Dynamics of Black-Tailed Prairie Dog Colonies Affected by Plague" (*Landscape Ecology* 23:255–67, 2008).

An early review of the impact of sylvatic plague on black-footed ferret recovery is presented by Mike Antolin and colleagues in "The Influence of Sylvatic Plague on North American Wildlife at the Landscape Level, with Special Emphasis on Black-Footed Ferret and Prairie Dog Conservation" (*Transactions of the North American Wildlife and Natural Resources Conference* 67:104–27, 2002). An early summary by Beth Williams and colleagues on the impact of sylvatic plague (and canine distemper) on the black-footed ferret decline at Meeteetse, Wyoming, can be found in "Disease and Endangered Species: The Black-Footed Ferret as a Recent Example" (*Conservation Biology* 2:66–74, 1988). Subsequent to that event at Meeteetse, plague remained an influence on black-footed ferret survival, as discussed in "Plague in a Black-Footed Ferret" (*Journal of Wildlife Diseases* 30:581–85, 1994). Most recently, Randy Matchett, who has overseen the ferret reintroduction at UL Bend since its inception, has published key findings on how enzootic (that is, non-epizootic) presence of sylvatic plague could be limiting ferret survival in "Enzootic Plague Reduces Black-Footed Survival in Montana" (*Vector-borne and Zoonotic Diseases* 10:27–35, 2010).

Details on the use of topical insecticides to control flea populations on prairie dog colonies and reduce the risk of plague outbreaks can be found in "Treatment of Black-Tailed Prairie Dog Burrows with Deltamethrin to Control Fleas and Plague" (*Journal of Medical Entomology* 40:718–22, 2003) and "Vector Control Improves Survival of Three Species of Prairie Dogs in Areas Considered Enzootic for Plague" (*Vector-borne and Zoonotic Diseases* 10:17–26, 2010).

Details on the plague vaccine used on black-footed ferrets can be found in "Vaccination with F1-V Fusion Protein Protects Black-Footed Ferrets against Plague upon Oral Challenge with *Yersinia pestis*" (*Journal of Wildlife Diseases* 44:930–37, 2008). An oral bait plague vaccine for prairie dogs is in development, and early advances by Tonie Rocke and colleagues can be found in "Consumption of Baits Containing Raccoon Pox-Based Plague Vaccines Protects Black-Tailed Prairie Dogs" (*Vector-borne and Zoonotic Diseases* 10:53–58, 2010) and "Sylvatic Plague Vaccine: A New Tool for Conservation of Threatened and Endangered Species?" (*EcoHealth* 9:243–50, 2012).

CHAPTER 12: KANSAS

An excellent popular article by Ted Williams on the struggle of Kansas ranchers Larry Haverfield and Gordon Barnhardt to conserve the prairie dogs on their land entitled "Doggone: Prairie Dogs Face a New and Deadly Threat" was published in *Audubon* magazine (November 2009). Jack Cully of Kansas State University is a nationally renowned prairie dog expert who has published widely on prairie dog ecology and plague. Some of his work specific to the state of Kansas is published in "Effects of Black-Tailed Prairie Dogs on Reptiles and

Amphibians in Kansas Shortgrass Prairie" (*Southwestern Naturalist* 46:171–77, 2001); "Ecosystem Engineering by a Colonial Rodent: How Prairie Dogs Structure Rodent Communities" (*Ecology* 89:3298–3305, 2008); and "New Records of Sylvatic Plague in Kansas" (*Journal of Wildlife Diseases* 36:389–92, 2000).

The work of writer Ed Abbey has become synonymous with the red rock deserts of southern Utah. Excellent examples of his work are the books *Desert Solitaire* (Touchstone, 1990) and *Beyond the Wall* (Holt, 1984). In addition to being known for its vast open spaces, the region has a high amount of biological diversity. The relatively high peaks and plateaus in the western part of the state in particular provide an elevational gradient that favors a high diversity of mammals, as detailed in Eric Rickart's article "Elevational Diversity Gradients, Biogeography and the Structure of Montane Mammal Communities in the Intermountain Region of North America" (*Global Ecology and Biogeography* 10:77–100, 2001).

The rarest of the five species of prairie dog, the Utah prairie dog, resides only in small populations in southwestern Utah. Because it is listed as a threatened species under the U.S. Endangered Species Act, a thorough review of the species can be found on the U.S. Fish and Wildlife Service's website www.fws.gov. For those interested in seeing a Utah prairie dog in person, the best opportunity is likely in Bryce Canyon National Park.

Results of field research discussed in this chapter involving the use and evaluation of systemic flea control products to mitigate plague risk can be found in "Field Evaluation of Imidacloprid as a Systemic Approach to Flea Control in Black-Tailed Prairie Dogs" (*Journal of Vector Ecology* 36:100–107, 2011) and "Mitigating Plague Risk in Utah Prairie Dogs: Evaluation of a Systemic Flea-Control Product" (*Wildlife Society Bulletin* 36:167–75, 2012).

Index